T0076105

WASTE MANAGEMENT

Selected Titles in ABC-CLIO's

CONTEMPORARY WORLD ISSUES

Series

Adoption, Barbara A. Moe
Campaign and Election Reform, Glenn H. Utter and Ruth Ann Strickland
Chemical and Biological Warfare, Al Mauroni
Childhood Sexual Abuse, Karen L. Kinnear
Corporate Crime, Richard D. Hartley
Domestic Violence, Margi Laird McCue
Emergency Management, Jeffrey B. Bumgarner
Energy Use Worldwide, Jaina L. Moan and Zachary A. Smith
Euthanasia, Jennifer Fecio McDougall and Martha Gorman
Food Safety, Nina E. Redman
Gangs, Karen L. Kinnear
Globalization, Justin Ervin and Zachary A. Smith
Illegal Immigration, Michael C. LeMay
Intellectual Property, Aaron Schwabach
Internet and Society, Bernadette H. Schell
Mainline Christians and U.S. Public Policy, Glenn H. Utter
Mental Health in America, Donna R. Kemp
Nuclear Weapons and Nonproliferation, Sarah J. Diehl and James Clay Moltz
Policing in America, Leonard A. Steverson
Renewable and Alternative Energy Resources, Zachary A. Smith and Katrina D. Taylor
Rich and Poor in America, Geoffrey Gilbert
Sentencing, Dean John Champion
U.S. Military Service, Cynthia A. Watson
U.S. National Security, Cynthia A. Watson
U.S. Social Security, Steven G. Livingston

For a complete list of titles in this series, please visit **www.abc-clio.com**.

Books in the Contemporary World Issues series address vital issues in today's society, such as genetic engineering, pollution, and biodiversity. Written by professional writers, scholars, and nonacademic experts, these books are authoritative, clearly written, up-to-date, and objective. They provide a good starting point for research by high school and college students, scholars, and general readers as well as by legislators, businesspeople, activists, and others.

Each book, carefully organized and easy to use, contains an overview of the subject, a detailed chronology, biographical sketches, facts and data and/or documents and other primary-source material, a directory of organizations and agencies, annotated lists of print and nonprint resources, and an index.

Readers of books in the Contemporary World Issues series will find the information they need to have a better understanding of the social, political, environmental, and economic issues facing the world today.

WASTE MANAGEMENT

A Reference Handbook

Jacqueline Vaughn

CONTEMPORARY WORLD ISSUES

Santa Barbara, California
Denver, Colorado
Oxford, England

Library of Congress Cataloging-in-Publication Data
Vaughn, Jacqueline.
Waste management : a reference handbook / Jacqueline Vaughn.
p. cm. — (Contemporary world issues)
Includes bibliographical references and index.
ISBN 978-1-59884-150-3 (alk. paper) — ISBN 978-1-59884-151-0 (e-book) 1. Refuse and refuse disposal. I. Title.

TD791.V38 2009
363.72'8—dc22

2008019139

13 12 11 10 9 1 2 3 4 5 6 7 8 9 10

ABC-CLIO, Inc.
130 Cremona Drive, P.O. Box 1911
Santa Barbara, California 93116–1911

This book is also available on the World Wide Web as an eBook.
Visit www.abc-clio.com for details.

This book is printed on acid-free paper ∞

Manufactured in the United States of America

For Terry Trumbull,
for introducing me to waste management

Contents

List of Tables

Preface

"Paper or Plastic?"

Since 1977, when the first plastic bags were introduced to the grocery industry as an alternative to paper sacks, consumers have had a choice about the type of waste they produce. After more than three decades of debate over which type of shopping bag is better for the environment, people are still not sure which one to choose. The more the population grows, the more goods that are consumed, and packaging, whether paper or plastic, becomes an important part of the waste stream—usually referred to as garbage or trash.

Waste, and how we manage it, typically ends up toward the bottom of the environmental policy agenda. Problems like air quality, global climate change, and energy are often the focus of policy makers' attention, with issues related to waste either ignored or deferred. But waste is actually an integral part of the overall environmental management picture. Increased consumer consumption of goods means more natural resources, such as forest products and fossil fuels, are used for packaging. More waste, in turn, means that more space is used for landfills or more particulate matter is emitted by incinerators. As the demand for electricity increases, nuclear power plants expand, creating more high-level radioactive waste. The paper versus plastic debate is used as a theme throughout the eight chapters of this book that follow. The controversy illustrates the shifting nature of the debate as technology advances, environmental consciousness is raised, population and consumption increase, and policies are created to mandate how waste is managed both in the United States and on a global level.

Chapter 1 provides an overview of the background and history of waste and explains that although trash has been a part of human existence since the very beginning of time, deciding what to do about it is a relatively recent phenomenon. The "problem" of waste focuses on the concerns raised by urban dwellers about disease and odors before the turn of the 20th century. The public health-oriented sanitarian movement and progressive reform (led in large part by women in major cities) brought attention to open-dumping practices, the use of swine as a refuse management tool as late as the 1960s, and use by residents of 55-gallon barrels as homemade backyard incinerators. The development of the sanitary landfill and the compaction of trash dates from the late 1930s and early 1940s. Chapter 1 also recounts the history of packaging consumer goods, one of the primary components of municipal solid waste.

Chapter 2 deals with current problems and controversies over waste management on the domestic level. The chapter continues with the issue of what type of container is "best" for consumers to use for carrying items such as groceries, and it introduces new issues such as the recycling of plastic, which often ends up in poor regions of China. Problems that have recently arisen, such as what to do with disaster debris from the attacks at the World Trade Center in 2001 or the ravaged remains of houses destroyed by Hurricane Katrina, are also explored. Among the solutions considered are new technologies such as phytoremediation to clean up waste sites and new uses for old waste sites as brownfield development.

The global perspective on waste management is covered in Chapter 3. In addition to discussing regional analyses of waste management, the chapter profiles the trash pickers who are a common sight in cities in developing countries, from Cairo to Buenos Aires. Segments are presented on efforts by groups such as Greenpeace to force Dow Chemical to clean up the toxic waste from the deadly gas leak in Bhopal, India, in 1984; on attempts to develop Australia's first national nuclear waste dump on aboriginal land; and on accusations that Japan has breached agreements on toxic waste trading among Asian countries.

Chapter 4 is an annotated chronology of events in waste management, beginning in the mid-18th century when Benjamin Franklin tried to organize workers to clean the city streets in Philadelphia. The chronology covers both domestic and international events. The individuals profiled in Chapter 5 are considerably less well known than the leaders who have sought to solve other environmental and social problems. Key figures in environmental history such as John Muir and Aldo Leopold were not unaware of the trash around them, but they focused on other issues related to wilderness, scenic preservation, and the use of natural resources, so they are not part of this chapter. Instead, included are brief snapshots of the lives of early refuse pioneers such as Jean Lacey Vincenz, who established the first sanitary landfill in the United States in Fresno, California, and sociologist Robert Bullard, who is widely regarded as the first scholar to develop the concept of environmental justice and hazardous waste facility siting.

Chapters 6, 7, and 8 provide various types of resources for the reader who wants to do further research and understand waste management as a policy issue. Chapter 6 includes facts, statistics, documents, and information that are not included in previous chapters. Included are materials that outline the different points of view, such as those surrounding the Yucca Mountain facility in Nevada and a list of the world's most polluted waste sites. Chapter 7 is an alphabetized, annotated directory of government agencies, both U.S. federal and state, that deal with the various types of waste, as well as nongovernmental organizations. The latter group includes industry trade associations, advocacy groups, and information clearinghouses that provide additional resources about waste. Chapter 8 concludes the book with a list of print and nonprint resources, organized by books, articles, and videotapes that are generally available. Finally, the book contains a glossary of key terms for those readers just developing an interest in and knowledge of waste management.

The goal of this book is to provide readers with background information and an overview of waste that has not been available in a single volume. The book represents an interdisciplinary approach combining history, political science, environmental science, engineering, sociology, and public health literature that weaves a story of how policy makers are finding new ways to deal with an old problem: trash.

1

Background and History

Most of us are introduced to the issues related to waste management as children when we are asked to carry out the family garbage. Usually, that garbage consists of refuse contained in a paper or plastic bag, sometimes used as a liner for a metal or plastic trash receptacle. According to the U.S. Environmental Protection Agency (EPA), more than a third of the waste generated in the United States comprises paper and product packaging. Sometimes, our trash includes items like aluminum cans, bottles, plastic milk jugs, and newspapers that we have put into a separate container or bin for curbside pickup and recycling by the trash collection service. Another 13 percent of municipal solid waste (MSW) comes from organic matter such as lawn clippings and food scraps (12 percent); rubber, leather, and textiles (7 percent); and wood (6 percent); with 3 percent of miscellaneous origin (U.S. EPA 2007a). But the journey of garbage from our homes to a place where it is buried in a landfill, incinerated, or recycled is a process that most of us know little about.

In the early history of humanity, containers for packaging were not required, as food was immediately consumed and human waste was just left on the ground to biodegrade naturally. Later, grass and reed fibers and clay were used to make pottery, which was in turn used to hold food and grains or items for barter and trade. Other natural products, such as hollowed-out gourds and animal hides, were used for such items as well. Pottery and similar vessels were reused and repaired until they wore out or were broken. Eventually, pit toilets were dug as primitive latrines for human waste, primarily because of unpleasant odors and concerns that menacing animals might follow the scent.

Paper was used as "flexible packaging" as early as the first and second centuries BCE, when the Chinese used treated mulberry bark to wrap foods. The technique of paper making advanced over the next 1,500 years, when the process was introduced to England in 1310 and to America in 1690. In those times, paper was made from flax and, later, linen rags; wood pulp was not used for paper until 1867 (Hook 2007). But paper was very scarce, and store owners asked their customers to bring their own tote bags—typically made of jute—when they shopped at the general store, where they could buy items from food and alcohol to sewing notions, clothing, and hardware. Little was wasted and almost nothing was thrown away. Parts from broken items were used to create other items or melted down.

The first paper bag was made in Bristol, England, in 1844, by the paper bag–making machine invented by Francis Wolle, who patented the process in 1852 and founded the Union Paper Bag Machine Company in Savannah, Georgia, in 1869 (Forest Preserve District of Cook County, Illinois 1971). A year later, Margaret Knight refined a machine to cut, fold, and paste square paper bag bottoms. Charles Stilwell was recognized for his improvement upon the paper bag bottom when he was issued a patent for a square bottom paper bag with pleated sides in 1883, called the S.O.S., or self-opening sack. It proved to be the model for today's paper bags, and its sales soared into the 1930s because the bags were strong and inexpensive to make.

The owner of a small grocery store in St. Paul, Minnesota, Walter H. Deubner, took the paper bag idea a bit further. He realized that his customers only purchased what they could carry home, and in an attempt to increase his business, he tried a new tactic to get them to buy more. He took the traditional paper bag and ran a cord through it to serve as a handle, which created a bag capable of carrying up to 75 pounds of groceries, and named the product the Deubner Shopping Bag. He sold each bag for 5 cents, he patented the method, and by 1915, he was selling more than a million shopping bags per year (Great Idea Finder 2007).

No one knows exactly who invented plastic bags; we do know that low-density polyethylene, the material of plastic, was invented in 1942. Canadian inventor Harry Wasylyk is credited with the invention of the green-colored garbage bag in 1950, when it was used for commercial use and first sold to Winnipeg General Hospital. Wasylyk and another Canadian inventor, Larry Hansen, sold the patent for the plastic garbage bag to Union

Carbide Company, which began selling garbage bags for home use in the late 1960s. Another small commercial version, this one made of clear plastic, was packaged and sold in paper boxes; these "baggies" began to appear in bakeries, in grocery stores, and at produce stands in 1957. Consumers began to use a different bag to separate each type of purchase they made, one for tomatoes, another for bread, still another for peaches, and so on. The baggies kept the produce from rolling around in the grocery cart, and because they were translucent, shoppers could see what they had selected, unlike the opaque paper bags they had used previously.

Although paper and plastic materials are the most commonly used for containers, metal and rigid metal packaging now comprises almost 8 percent of the municipal waste stream. Although tin plating was discovered in Bohemia in 1200 CE and used to make cans in the 14th century, metal was not used for food storage until the early 1800s; metal containers were used to sell snuff in pre–Revolutionary War America, and in 1809, Emperor Napoleon Bonaparte of France offered a large reward to anyone who could successfully preserve food for his armies. A year later, a British inventor, Peter Durand, received a patent for a sealed cylindrical can. But today's model of aluminum cans did not appear until 1959, about the same time as did the pop-top lid for cans. The pop-top was used initially for beer and soda and in the 1980s and 1990s was expanded for use with items such as potato chips and peanuts. In response to complaints from consumers about the need for a readily available can opener, the pop-top lid could be used by anyone, even children, to gain access to a product. Campers could open up a can of ready-to-eat spaghetti and meatballs, and the idea of opening up a beer with the whooshing sound of the air being released was popular. For older consumers with less strength in their hands and for those living alone, pop-top cans and their single-serving-size packaging make meals of canned soup, tomato sauce, and fruit cocktail easier to prepare.

Glass, which makes up about 5 percent of the waste stream, was industrialized in Egypt about 1500 BCE and made from raw materials such as silica and soda ash that are combined and heated and then molded into shapes for glasses, bowls, and containers for storage. Over the next millennium, processing methods changed to allow the production of bottles, with the first automatic rotary bottle-making machine patented in 1889. Glass was sturdier than other types of packaging and could be reused and refilled, making it cost effective to produce.

The most recent major addition to the waste stream is plastics, which make up nearly 12 percent of municipal solid waste. Originally used mostly for military purposes, plastics were discovered in the 19th century but not refined until 1933 by German manufactures. Various forms of plastic, such as celluloid, vinyl chloride, and foam, were developed for commercial uses by the mid-1950s, although some plastic compounds were reserved purely for military applications. Today's most common plastic container is made of polyethylene terephthalate, or PET, a type of resin. The beverage container industry introduced PET to the consumer market in 1977, and other industries began packaging their products in PET in abundance by 1980.

To make them more lightweight, less expensive to produce, and reusable, most plastics are now designed to be disposable, meaning that they are used once and then thrown away. Their disposable nature is one of the major reasons why industrialized countries are dealing more with waste management problems than they did in the 19th and 20th centuries; more products are available for purchase, and they are wrapped in containers and packaging that are disposable and of little value to consumers, which means more refuse ends up in the waste stream, needing to be disposed of in one of several different ways described later in this chapter.

Problems in Defining and Measuring Waste

As the categories below indicate, waste can be identified in many ways. Various disciplines define the term *waste* differently, so what a public official may call a certain type of waste may be different from the perspective of an engineer or city planner. Even more important, questions have emerged about the accuracy of statistics that show how much waste is both generated and disposed of each year. Some state regulations, for instance, include materials such as construction debris under the category of municipal solid waste, while in other states, that debris might be placed in a separate category. Some states consider medical waste and contaminated soil as hazardous waste; others use terms like *biohazard* for these materials. Some wastes that are managed at the site where they are generated (such as a manufacturing facility) are not considered part of the total because they never go through the collection and disposal process. The lack of agreement about definitions and lack of accurate data make it difficult for policy managers to determine how serious the problem might be and how best to deal with it.

Local, state, and federal legislation and regulations require leaders to know what is in the waste stream, the costs of collecting and disposing of those wastes, and any health and safety issues that might be raised. However, concerns about waste management tend to fall toward the bottom end of the policy agenda and are less visible than other environmental issues such as water or air pollution. Waste management becomes important to many policy makers only when health or environmental issues reach the crisis stage.

Types and Sources of Waste

The term *community waste* is sometimes used interchangeably with *municipal solid waste,* although there is no standardized or accepted definition of what types of waste fit into which categories. But generally, the types and sources of waste are considered to be part of the flow of materials in society and a function of land use, zoning, and consumption. The following categories are those that are usually recognized by both policy makers and waste managers.

Household Waste

The unwanted materials produced by single- and multifamily homes, apartment complexes, and other dwellings consist of food and other types of organic waste; paper products and cardboard; plastics; textiles; leather; yard clippings; wood; glass; metal; ashes; and various bulky items such as tires, televisions, furniture, and appliances. The latter group of products may also be called *bulky waste* or *white waste.*

Commercial Waste

Retail establishments such as restaurants, stores, hotels and motels, office buildings, and similar businesses create commercial waste, which includes food scraps, paper and cardboard, glass, metal, plastic, wood, and other items similar to those produced by households. These wastes are often commingled with residential waste products when they are collected, although the costs of collecting commercial waste may be higher.

E-Waste

Waste from electronics such as computers, cell phones, and televisions is a growing concern because of the short life span of the products and the lack of consumer knowledge about what to do with them when they become obsolete or stop functioning. In the past, most items were simply tossed into the regular trash without any consideration for the materials from which they are made. But these products' components, ranging from lead to cadmium to mercury to various plastics used in circuit boards, cables, covers, and connectors, pose a number of risks. The waste is not suitable for incineration because some components form toxic substances when burned. When placed in landfills, they can leach into the soil. The issue of how best to handle this specialized type of waste is outlined in Chapter 2.

Hazardous Waste

The EPA has defined waste as hazardous if it exhibits at least one of four characteristics: ignitability, reactivity, corrosivity, or toxicity. Sometimes, these wastes are listed by name, although the list is not exhaustive, and sometimes, they are regulated by several agencies. In other cases, consensus has not been reached among agencies as to whether or not a waste is hazardous. Hazardous waste is a problem not only because of those four characteristics but also because of other factors such as the quantity of waste and its mobility. Today, waste managers also look at the types of sites that receive specific hazardous waste, such as their proximity to human populations, containment integrity, local groundwater hydrology, potential impact on flora and fauna through transmission, geological integrity, and the ultimate level of controlling the waste. The more that is known about the characteristics of the waste, the better informed decision makers can be about how to treat it. In areas and at times where awareness is lacking about the negative impact of hazardous waste, as was the case in the United States in the 1940s and 1950s, little motivation exists for implementing expensive treatment options or protecting workers and the public. These issues are covered in chapters 2 and 3.

Industrial Waste

Industrial waste emanates from trade processes, such as the making of steel and automobiles, the building and construction industry, coal mining, food processing, metal finishing, petroleum refining, and utility power plant operation. During the first half of the 20th century, the creation of industrial waste was considered part of the cost of economic growth and development. In 1914, George Price suggested that noxious industrial waste could be controlled in two ways: removing and destroying it or storing it and related materials in closed and tight vessels. Another recommendation was that industrial waste be isolated by segregating it at remote and public locations. Engineered solutions included using metal fencing material to contain industrial waste (Colten and Skinner 1996, 13). Generally, industrial waste became an issue much later than that of organic material and other forms of refuse. Because it was assumed that industrial waste contains no bacteria and therefore poses little threat of disease, many substances were routinely dumped into watercourses or sewers. This theory proved to be erroneous. Other assumptions included the belief that the components of industrial waste acted as effective bactericides, another misconception.

Building Waste

Construction and demolition create waste when roads are built, houses are torn down, pavement is repaired, and new developments are created in communities. These wastes include dirt, wood, steel, concrete, asphalt, and other materials used in building and renovation. These materials also become debris, which must also be dealt with when accidents, emergencies, and natural or human-caused disasters occur, such as a major earthquake in which buildings are destroyed. Disaster debris is discussed further in Chapter 2.

Medical Waste

Medical waste (which includes wastes generated from human activity) is important because it involves multiple federal, state, and local agencies; laboratories; blood suppliers; universities conducting research and patient care; transporters and disposal firms; and medical facilities ranging from doctors' offices to hospitals. In 1968, the U.S. Public Health Service divided hospital waste into six categories, warning that legal implications were attached to the handling of these types of waste and that precautions should be taken to ensure the safety of staff, collectors, and the public (Turnberg 1996). The EPA provided guidelines for hospital waste in 1974 and advised medical providers of the difficulty of segregating these wastes from other nonhazardous wastes, so all hospital waste must be considered potentially contaminated and must receive special handling and treatment. The EPA's *Guide for Infectious Waste,* published in 1982, recommends six categories of medical waste for special management, along with specific categories for consideration as infectious waste. After years of study, the EPA has now established definitions in the medical waste category to differentiate medical waste that consists of an infectious substance, termed *regulated medical waste,* from *regular medical waste.* The differentiation reflects the need to treat medical waste containing an infectious substance differently from regular medical waste in treatment, packaging, and transportation. The former type of waste requires special procedures for handling that are consistent with requirements established by the United Nations' World Health Organization in 1999 (WHO 1999). Regulated medical waste is now defined as a waste or reusable material known to contain or suspected of containing an infectious substance and generated in the diagnosis, treatment, or immunization of animals; research on the diagnosis, treatment, or immunization of animals; or production or testing of biological products.

Agricultural Waste

Crops, orchards, vineyards, livestock operations, and farms create waste when food spoils, trees are pruned, fields are cultivated and produce is picked, animals are fed and produce manure, and chemicals are used. This category, agricultural waste, is also known as biomass; it differs from similar types of waste that are generated by households. Most agricultural waste is handled on-site, such as burning a field after a crop has been harvested.

Woody Biomass

Logging and milling operations produce organic materials that can be used for the production of energy. This category includes tree limbs, slash piles, sawdust, needles and tops, and small-diameter trees that are unmerchantable and are removed as a part of forest-thinning operations. A 2005 study conducted by the U.S. Department of Agriculture and the U.S. Department of Energy estimated that more than 369 million tons of woody biomass are available for energy use in the United States every year (Bryant 2007).

Universal Waste

In 1995, the EPA finalized regulations for universal waste (UW), a descriptive term applied to wastes that are widely generated by both businesses and households. The new regulations allowed a wider range of wastes to be managed under less stringent requirements than under hazardous waste rules. The four categories of UW include unused hazardous waste pesticides that have been recalled under federal laws or collected in waste pesticide programs; mercury-containing thermostats; lamps; and spent batteries, including those made of nickel-cadmium and lead-acid alloys.

The regulations were created for two reasons. First, the EPA wanted to streamline the handling and management of these products, which previously were considered only as hazardous waste. Requirements for labeling, employee training, shipping, and tracking were changed by the regulations, and businesses were allowed to accumulate and store materials on-site, relieving them of having to hire a hazardous waste shipper to handle the materials. Second, the new definition, adopted in part because of requests by both industry groups and environmental organizations, was designed to encourage retailers and other entities to collect these types of items for recycling or treatment. This new definition of waste covers widely generated materials and reduces the quantity that would otherwise end up in the municipal waste stream and perhaps in landfills. States are permitted to create additional categories of universal waste if they choose to do so. While individual households are encouraged to participate in UW recycling programs, they are not required to do so, and the items can still be included in household trash.

Waste Management Options

State and local governments can choose from a limited suite of technologies and programs to deal with waste, from simply collecting trash and dumping it to highly technical processes that are being used in pilot studies to determine their effectiveness. This section identifies the options that are available, in alphabetical order.

Collection

Most communities rely on compaction trucks or other types of vehicles to bring waste from the location where it is created or produced to another site for sorting and treatment. Collection is the most expensive element of waste management, accounting for between 50 and 70 percent of the costs for operations. These costs include those associated with hauling the waste to a disposal location, where it may be dumped or processed. Some of the wastes are commingled and may not be separated out by type. In most communities, residential trash is collected by large vehicles that pick up waste that has been placed in some type of bin. Trash trucks cost from $100,000 to $150,000 each and may be operated with a crew of from one to four people, depending on the type of vehicle used. Some waste is noncontainerized and is picked up by a different type of vehicle with a claw-like apparatus designed to pick up bulky items such as tree limbs. Collection may be a function of a municipal government as part of the services it provides or it may be a service provided by a private company that is contracted to collect the waste.

Composting

Composting is a method of dealing with organic waste and can be done by both individuals and municipalities. Composting is considered a valuable strategy because the volume of waste can be reduced by as much as half by breaking down easily degradable plant and animal tissue, such as yard waste and food scraps. The most common type of composting involves seasonal leaf collection and Christmas tree pickups. Some composting is done through the use of sheltered or unsheltered windrows, which are periodically turned to speed decomposition, or through in-vessel systems whereby the material is placed in an enclosed drum, tank, or silo in which conditions to maximize the breakdown of nutrients can be closely controlled.

A number of considerations may limit composting on a large scale, but they primarily involve odor and fly and rodent attraction, consequences of composting that cause problems for waste managers and residents alike. Odors are not normally a health concern, but they do affect the quality of life in a community where a facility is located or where outdoor transfers and composting occur. Enclosed facilities can help to eliminate the odor issue, but they are seldom cost effective. Organic residues inevitably attract vectors (insects and animals that may transmit diseases), especially food waste and sewage sludge; enclosure of the facility does reduce this problem but is costly. An additional issue is making sure that no hazardous substances, such as pesticides, are present in the compost when it is initially collected. The problem occurs when humans or animals eat food that was grown on ground where compost is deposited after processing. Heavy metals can pass through the food chain, as can pathogens. The problem is less apparent when the compost program involves only yard waste and the products are used primarily for landscaping (mulch) or in industrial boilers as fuel.

Incineration

One of the common ways in which the volume of waste is reduced is through combustion or incineration, a controlled burning process used by both governments and private companies around the world. Some facilities convert the heat with water to produce steam, which can then be used to generate electricity, called *waste-to-energy*. In recent years, operators started separating the recyclable components of waste, such as glass and metals, and then shredding the remaining materials before burning. The process is used by more than one-fifth of U.S. municipal solid waste incinerators to produce refuse-derived fuel, which can then be used in facilities such as power plants.

Incineration works best for nonhazardous waste, because destroying chemical compounds and disease-causing organisms is sometimes difficult, even when burning material at very high temperatures. The residual ash that is created, consisting of the elements originally present in the waste, can often be used for beneficial purposes, such as in constructing roads or building parking lots. The EPA samples the ash to determine whether or not it contains any traces of hazardous material; if so, it must be treated as hazardous waste with special handling procedures.

While some consider incineration a renewable power resource because it burns material, such as wood, paper, tires, food waste, and yard clippings, that would otherwise be sent to landfills, it does have drawbacks. For instance, the process of incinerating municipal solid waste creates pollutants that are controlled by federal Clean Air Act regulations. Facilities must secure emissions permits because, depending on what is being burned, they may emit nitrogen oxide, carbon dioxide, sulfur dioxide, and trace amounts of toxic pollutants such as dioxin.

Integrated Waste Management

Some states have adopted a process for selecting which types of techniques, technologies, and management programs they should use for achieving waste management objectives, including four methods identified by the EPA: source reduction, recycling and composting, combustion, and landfills. Many states and communities then prioritize these options in order of preference because of the environmental impacts of each one. California, for instance, has chosen a hierarchical process that gives preference to source reduction, followed by recycling and composting, waste transformation, and landfilling. Combustion has been replaced by waste transformation as an option in that state, although the process still requires that recycling be considered only after the maximum amount of source reduction has been achieved (Tchobanoglous and Kreith 2002, 1.8).

Landfills

Early landfills, or dumps, were simply open trenches, sometimes called tips, in which trash was heaped into piles and sometimes covered with 12 to 15 inches of soil and then layered with other refuse such as ash. Several large urban communities, including Seattle, New Orleans, and San Francisco, built landfills during the early 1900s, and the U.S. Army Corps of Engineers also experimented with landfills on military bases.

Today's landfills are totally different from those built a hundred years ago. Technology and sophisticated design are now applied to make landfills one of the most frequently used ways of dealing with waste. The term *sanitary landfill* originally applied to facilities where waste was covered at the end of each day's operation, but today, it refers to an engineered facility for municipal solid waste operated to minimize environmental and public health concerns. A *secure landfill* is one specializing in the management of hazardous waste (Tchobanoglous and Kreith 2002, 14.2).

In most facilities, a deep trench, canyon, or depression with an earth embankment is excavated. Sometimes these sites are naturally occurring, such as an old quarry, but they are usually purpose built for the location. Landfills are now usually built with a flexible liner, made of clay or a synthetic material, that is used to cover the bottom and sides of the facility. The liner helps to keep any gases or liquid (known as leachate) from migrating out of the landfill. Some materials may be banned from landfills, including paints and solvents, motor oil, batteries, and pesticides, unless the facility is permitted to handle these types of products using specialized treatment (U.S. EPA 2007b).

Mechanical Biological Treatment

Waste management by way of mechanical biological treatment involves both the mechanical sorting of waste and the biological treatment of organic waste. In facilities where this strategy is used, a bulk handling machine sorts out the recyclable components of garbage, and that portion which is biodegradable is then broken down through anaerobic digestion or composting. Some of these treatments produce biogas, used to generate energy, and others produce soil amendments that may be used by consumers to augment poor soil conditions.

Phytoremediation

In the early 1990s, researchers began testing the use of green plants and their associated microorganisms to stabilize or reduce the contamination in waste soils, sludges, and other materials. Since then, the process, known as phytoremediation, has been used at more than 200 hazardous waste sites. It manages waste that does not involve human-made chemicals, and it works especially well on sites with low concentrations of contaminants over large cleanup areas. The plants take up contaminated waste through their root systems, along with water that may or may not be contaminated. Along with native plant species, trees such as poplars and plants such as sunflowers are among those used because they grow quickly. These plants can adapt to local climates, are easy to plant and maintain, and have the ability to take up large quantities of water through their roots, facilitating the process (U.S. EPA 2007c).

Recycling

Recycling generally refers to the separation of materials in the waste stream so that some of those materials can be reused. Recycling is highly dependent on two factors: the availability of recycled goods supplies and the market for those goods. From 1994 to 1997, for instance, the United States experienced a glut of paper that had been collected for recycling; the market for old newspapers dropped off, as did prices, and some communities even had to store paper in warehouses as they waited for the prices to rise again. Recycling is considered one of the best types of waste management because it can stretch the capacity and life of landfills by reducing the volume of materials considered as trash, while reducing the amount of virgin material that is needed for products and packaging. According to the Bureau of International Recycling, without recycling, a substantial number of end-of-life goods would end up in landfills, representing an environmental and economic loss because the materials they contain would be lost to the production cycle forever (Bureau of International Recycling 2007).

Three types of recycling methods are used:

- Source separation: Households and businesses separate their recyclable items from other waste at the source where they are created, placing material into small containers or bins by type (paper, glass, plastic). Processing is usually conducted by scrap dealers or taken to a consolidation site for sale to another processor.
- Container separation: Large bins or containers are used to separate waste into three categories: nonrecyclable waste, organic waste, and recyclables. Once at the collection facility, the recyclables are separated out by type, sometimes by manual sorting. Some communities rely on a bag system, with all recyclables placed in a visually distinct blue bag for collection (blue bags differentiate recyclables from trash and yard waste, the bags for which are usually black or green).

- Commingled collection: All waste is commingled and collected by a single truck. Recyclable materials are sorted and processed by mechanical means including magnets and shredders. This method is sometimes called front-end processing or refuse-derived fuel processing.

Most recycling efforts focus on homeowner or community involvement, such as the provision of bins by the local government to home dwellers so that items can be picked up at the curbside as a regular part of trash collection services. Other areas have experimented with drop-off stations for items such as paint, appliances, consumer electronics, and Christmas trees, which are then recycled and processed depending on the type of material. Another strategy is a buy-back program, a low-cost, centralized operation whereby items such as aluminum cans or glass bottles can be given to a recycler for a fixed amount of payment, such as by the pound. This method is most successful in states that have bottle deposit laws, which encourage recycling and reduce the amount of waste that must be landfilled or incinerated.

Source Reduction

One of the more recent strategies is to reduce the volume of waste, or alternatively, to reduce the level of waste toxicity. A variety of ways exist in which source reduction, also known as pollution prevention or waste prevention, can be accomplished, such as designing packaging in such a way that the product can be refilled, as, for example, with shampoo. Maintaining and repairing durable products, such as bicycle and automobile tires, can increase their use and alleviate the need for more frequent replacement. Other methods of source reduction include developing more concentrated products, such as laundry detergent, or packaging products in larger bulk sizes. Much of the emphasis in source reduction is placed on consumer choice—making buyers take responsibility for the amount of waste they produce. For example, offices and universities might reduce paper waste by telling their employees or students to print documents on both sides of paper; electronic messaging can reduce paper consumption even more. Providing linen napkins in a restaurant cuts down on the use and disposal of paper napkins. Source reduction may also require a change in behavior that cannot be legislated or regulated by government, only by societal or peer pressure.

Consumers might be encouraged to reuse containers rather than throwing them away and buying new ones, to rent or borrow supplies, and to share items such as catalogs and magazines. Other techniques are to request being taken off of bulk mailing lists, receive statements and bills electronically rather than by paper copies, and donate unwanted items to charities.

Not every strategy can be used in every situation, however. Urban dwellers may find it difficult to set up a backyard composting bin, and some documents must be printed on paper for legal reasons. Garage and yard sales are prohibited in some residential areas, and not every community has set up a hazardous waste collection program. But these forms of resourceful living allow for a mixture of programs and ideas that can be adapted to a household or commercial operation rather than requiring a one-size-fits-all approach.

Waste Transfer Stations

In many communities, both urban and rural, municipal solid waste goes through another processing point called a waste transfer station. A garbage collection truck may take its load to one of these facilities, where the waste is stored temporarily until it can be loaded onto a larger truck or transport vehicle for shipment to a disposal site, such as a landfill. Using a transfer station can reduce the number of vehicle miles traveled to and from the landfill, allowing smaller trucks to make shorter trips or maneuver through urban streets more easily than large vehicles. Transfer stations can also be located outside of residential areas to avoid the problems of odor and flyaway waste that become a neighborhood nuisance. Waste transfer stations may also include loading docks, parking areas, or storage facilities.

History of Waste Management

Any history of waste management in the United States begins in colonial America, where farmers simply cast off organic waste resulting from agricultural production. Recycling was a common element of a society in which manufactured goods were rare and costly. If something was broken, it was not tossed into a waste bin but was reconfigured in some way to make it useable for some other purpose. Old, tattered clothes, for instance, were ripped into long strips and used as bandages, for curling hair, or for tying up garden vegetables on poles. Scrap dealers traveled the countryside buying up unwanted or broken items that could later be sold to someone else who found a use for them.

As urban centers began to develop in the 18th and early 19th centuries, the waste stream consisted primarily of animal manure, food scraps, and human waste. By the mid-19th century, those wastes (along with fireplace ash and street sweepings) were used as fertilizer on fields or as a soil amendment. Human waste, also called night soil, was picked up by laborers paid by households wealthy enough to hire someone else to take care of an odorous problem. Even small communities had crews to clean up horse manure on streets and thoroughfares, since the animals' owners took no responsibility for the waste they created. "Refuse is primarily an urban blight," notes historian Martin Melosi. "Agrarian societies generally avoided refuse pollution, while cities and towns faced the gravest dangers" (2001, 93).

The Industrial Revolution led to wondrous inventions and discoveries, but the affluence of the United States also led to an increasingly serious problem in the form of solid waste. During most of the 19th century, municipalities focused primarily on the collection of food waste and horse manure. Horse-drawn carts were used to gather waste from piles in the street, with putrefaction in trenches and piles the primary method of disposal. In some areas, refuse was simply dumped into the ocean or other bodies of water. Organic materials, once disposed of, were treated with the naive attitude of "out of sight, out of mind." Melosi notes, "As more was learned over time about the changing nature of the waste stream, older collection and disposal methods came into question and were sometimes modified, but they were rarely abandoned" (2001, 71).

In many cities, scavenging through refuse bins and piles of garbage on the streets was a way to make a living. Cart men, scow trimmers, ragpickers, and urban residents were often exposed to filth and disease, and reformers attempted to clean up urban centers and slaughterhouses as part of the Progressive Era at the turn of the 20th century. Groups like the Citizens Association of New York, the Civic Improvement League of St. Louis, and the Philadelphia Municipal Association organized not only to clean up the trash but also to seek ways of ensuring that it was treated, despite some unscrupulous city contractors who sometimes just dumped barge loads of refuse into rivers and lakes (Rogers 2005).

Documentary film producer and author Heather Rogers makes an interesting observation about the development of waste management as a social and political problem:

> Garbage as we know it is a relatively new invention predicated on the monumental technological and social changes wrought by industrialization. Until mass production became the norm in the United States, manufactured commodities were always expensive and not always available; most ready-mades were imported from Europe. Such items were too dear to use once, then discard. The waste of pre-industrial and early industrial societies was comparatively minimal and for the most part could be absorbed back into the earth.
>
> That's not to say there weren't serious health problems due to a lack of organized refuse management in cities, but the *contents* of the rubbish bin were relatively benign. (2005, 31)

George E. Waring, a Civil War veteran who became commissioner of the New York Street Cleaning Department in 1895, was the first U.S. sanitarian, responsible for setting up a municipal trash collection and street-cleaning office for the city. Waring, who is profiled in Chapter 5, pioneered new methods of both collecting and treating municipal waste, and he developed a corps of workers known as the White Wings because of their starched white uniforms. Waring put an end to the scavenging and illegal dumping of waste by developing a system whereby residents presorted their trash and licensed haulers loaded it onto barges, which were taken to a small island for processing.

One popular method for dealing with waste, the incinerator or combustor, was brought to the United States from Europe in the late 19th century. The first garbage incinerator was built by the U.S. Army on Governor's Island, New York, in 1885 to cremate solid waste; an estimated 180 additional incinerators were constructed over the next 25 years. By 1910, however, incineration was no longer considered a waste panacea because American garbage tended to be wetter (meaning that it contained more organic material such as food scraps) than European trash, and it took much more fuel to burn. The facilities were capital intensive and relied on skilled labor, making incineration more expensive than other reliable methods for disposing of waste.

Waring's reforms in waste disposal led to the growth of the public health movement in the United States. The discovery of bacteria as vectors for disease and the spread of virulent illnesses in Europe, and urban population growth in American cities led to the development of urban sanitation departments and academic programs focusing on waste disposal. Municipal governments dealt with both the collection and the disposal of waste, recognizing that what was once merely a nuisance was now a serious problem that had been neglected for too long. Public works departments took over the task of dealing with household refuse, some of which was collected by contractors, and lucrative agreements emerged at the same time as complaints about corruption among city officials. By the early 20th century, local governments also began separating street cleaning, which was easy and cheap, from garbage collection, which was expensive and complex and considered a more technical problem as sanitation engineering emerged as a profession. The first organized curriculum in sanitary engineering was offered at the Massachusetts Institute of Technology in 1894.

Waterway dumping became an expedient method of dealing with garbage, especially for communities adjacent to rivers and along coastlines before the turn of the 20th century. In Seattle, for instance, residents complained about poor trash management in the 1890s, so city officials began off-loading garbage and ash into Puget Sound. Chicago trash was dumped into Lake Michigan, even though the lake also served as the city's water supply and commercial fishery. Louisiana officials justified the practice of dumping garbage into the Mississippi River by rationalizing that dumping trash into such an immense body of water that was in constant motion would result in only a limited impact (Rogers 2005, 67–68).

Another popular method for getting rid of garbage was feeding it to swine. Organic waste was commonly separated out from other types of trash, and hog feeding was even recommended as a food producing and waste conservation method by the U.S. Food and Drug Administration during World War I. Major swine feedlots, sometimes called piggeries, were built in cities such as Los Angeles and Kansas City, Missouri, and in the 1930s, feeding swine organic refuse was the most prevalent method of food waste disposal. By 1939, the U.S. Public Health Service estimated that 52 percent of cities were feeding garbage to swine.

With industrialization, the American Water Works Association focused on a different problem in the 1920s, claiming that most industrial wastes were detrimental to water supplies. Consumers were aware that industrial waste contaminated local waterways, imparting an offensive taste, discoloration, and odor, but they knew little about the issue beyond turbidity, or cloudiness of the water. Gradually, incidents throughout the United States focused attention on the health problems. In the early 1930s, public health officials found that a municipal garbage dump near Indianapolis had tainted local water supplies. Similarly, Southern California officials began monitoring groundwater supplies, looking for industrial waste contamination. In the "Montebello Incident" of 1945, they found evidence of an organic chemical from a weed killer manufacturing facility that had traveled more than 15 miles to contaminate public water wells. As the incidence of contamination increased, public health officials took a second look at using the ground to dump waste (Colten and Skinner 1996, 24–25).

Prior to the 1930s, only a few industries, such as smelters and some refineries, created waste that was considered dangerous, and even those produced relatively small volumes. As the chemical industry grew during the 1930s and 1940s, the amount of waste produced grew astronomically, especially synthetic organic chemicals used as pesticides and fungicides. Added to the mix were chemical warfare agents. Disposal of the wastes ranged from incineration to burial in landfills, open dumping, and depositing in waterways and quarries. Short-term considerations appeared to determine what type of disposal method was used, and where (Colten and Skinner 1996, 5–7).

Waste management was almost always considered a local government problem, and between World War I and World War II, municipalities were usually able to handle the trash generated by residents. Gradually, however, the amount of solid waste began to increase; the National League of Cities estimated that the rate of increase after World War I was five times the rate of population growth (Luton 1996, 122). Cities began operating sewage treatment plants during the 1930s as they sought ways to differentiate among the various types of waste and how to treat them. Prior to World War II, however, the practice of separating organic materials and putting them back into the soil was abandoned, as was the idea that waste services should pay for themselves through salvaged materials. Trash was seen as material that no longer had use value. In the hands of the sanitation engineer, garbage was handled as an innocuous and unavoidable class of debris. But by the 1930s, all U.S. cities with populations of more than 100,000 adopted some form of organized refuse collection and disposal (Rogers 2005).

Sanitary landfills were almost universally praised as the solution for garbage and municipal waste through the 1940s, although professional organizations noted the need for safeguards. Dumps, which were considered the least expensive form of disposal for nonorganic garbage, were no longer isolated in the countryside, but instead were purpose-built sanitary facilities that made it less expensive to transport trash away from the city. The first landfill was built outside the city limits of Fresno, California, in 1934 using compaction and then covering the trash with dirt to keep rodents out. Ironically, the facility would later be added to the nation's Superfund list (see below) as a hazardous waste site needing remediation and cleanup.

Over the next 25 years, landfill operations were improved to maximize compaction and operate with fewer workers, becoming more cost effective than incineration; by the mid-1960s, only about 10 percent of solid waste in the United States was being incinerated, with the majority of the remainder buried in landfills. But as environmental concerns surrounding landfills began to grow, pressure was placed on the federal government to examine the consequences of using landfills, such as groundwater contamination, harmful air emissions (especially with the growth in the plastics industry and products), and odor control. The emerging environmental movement provided better solutions to dealing with trash, which had not only become a visual blight but also had a negative impact on the quality of life.

Those concerns were exacerbated by the growth of disposable packaging that became the norm after World War II. Procter and Gamble introduced the first disposable diaper in 1961, and new products like disposable razors and single-use aluminum containers that could be thrown away became commonplace. Throwaway containers and attractive packaging increased the amount of trash that needed to be disposed of, especially synthetics. By 1960, plastics surpassed aluminum as one of the largest industries in the United States, but also seen was an increase in toxins entering the waste stream. Polyethylene containers, a major source of toxic materials, were used for nearly every conceivable consumer need. In the 1960s, there were concerns that some plastics might potentially be dangerous, especially if burned, producing toxic fumes. The debate about the safety of plastic continues, with some studies warning of potential health risks.

Consumers gravitated toward purchasing beverages in containers made of polyethylene terephthalate. Bottled water sales, for instance, increased from 3.8 billion units sold in 1997 to 27.9 billion in 2005. These bottles frequently end up as litter and have only a 23 percent recycling rate, lower than any of the other most common packaging materials. This becomes a waste management problem because the bottles that do end up in the trash do not degrade. The Recycling Institute estimates that about 18 billion barrels of crude oil equivalent were consumed in 2005 to replace the 2 million tons of PET bottles that were wasted instead of recycled (Groff 2007). The problem was exacerbated as the era of refillable glass bottles ended, an issue described in more detail in Chapter 2.

By the turn of the 21st century, waste management was almost a nonissue, compared with other environmental problems. Global climate change, wildfires, droughts, drinking water shortages, and endangered species crowded virtually everything else off the political agenda. Occasionally, a news report would cover a new development at the Yucca Mountain nuclear waste repository, or a press release was featured about the 10 most polluted places in the world (detailed in Chapter 6). But scholarly articles and books and government funds for waste researchers appear to have dried up, along with the public's perception of urgency. Dealing with garbage became an issue no one seemed to think about; much the same way it was out of sight, out of mind earlier in U.S. history.

Federal Legislation and Regulation

In reviewing federal waste legislation from a chronological perspective, it might seem as if the U.S. Congress and the executive branch of the federal government moved back and forth from one type of waste or management strategy to another, and indeed, this is largely the case. During the early 20th century, for instance, the federal government began to regulate some types of materials while seemingly ignoring others. In 1908, the transport of hazardous materials and waste was regulated by the Explosives and Combustibles Act, which focused on the rail transport of explosives used in mining. Officials agreed that certain operations that used hazardous materials needed to be monitored so that the waste from those operations was handled properly. But since waste management was considered primarily a local problem, the federal government seldom stepped in unless the issue involved national security or interstate commerce.

In contrast, radioactive waste developed as an issue in the 1950s and 1960s as government officials attempted to find ways to deal with the accumulation of materials that were the by-products of the military's buildup of weapons after World War II and during the Cold War with the Soviet Union. The focus was on national security and the nonproliferation of nuclear technology and materials to other countries. After passage of the Atomic Energy Act in 1954, businesses partnered with the government to commercialize nuclear power plants, although the amount of waste from these facilities was only about one-quarter of the waste produced from weapons research and production. The Atomic Energy Commission (AEC) realized that eventually some sort of large storage alternative would need to be developed, and in 1955, it asked the National Academy of Sciences to explore options for disposal. In 1957, the Academy issued a report recommending that the most promising method for disposal would be burial in underground salt mines through which no water could pass; such mines are located on the south rims of the Great Lakes, which range from New York to Michigan (McCutcheon 2002).

The government, however, chose a salt mine near Lyons, Kansas, to begin studies of radioactive waste disposal, spending 10 years in research there before abandoning the site as unacceptable in 1972. Residents and political leaders in Carlsbad, New Mexico, decided that they had a perfect site 26 miles east of the city and heavily lobbied state and federal officials to consider the location for a repository. Their motivation was almost purely economic, as they sought a replacement for the jobs lost in the declining potash mines. The AEC accepted the proposal and announced plans to study the site as a potential $25 million nuclear waste storage facility. In 1979, Congress authorized the Waste Isolation Pilot Plant (WIPP) as a repository for disposing of radioactive waste resulting from defense activities.

Thus began a political battle between environmental groups and critics of the plan that delayed the facility's opening for more than a decade as officials tried to determine whether safety and other concerns were being adequately addressed. It was not until 1991 that the secretary of the Department of Energy informed the secretary of the Interior that the WIPP was ready to begin accepting wastes for a five-year test period. President George H. W. Bush then named the EPA as the WIPP's independent regulator, setting the stage for the development of operating criteria. In 1998, the EPA announced its certification of compliance with the criteria, and a year later the first shipments of nuclear waste were sent from New Mexico's Los Alamos National Laboratory.

The federal government's role in managing other forms of waste has shifted from one that was purely a local function initially to one that requires the participation of public entities at all levels. Statutory control began in 1965 with the Solid Waste Disposal Act (SWDA), which was enacted as part of the Clean Air Act amendments. Congress realized that state and local governments were no longer capable of handling waste on their own as the nature of the waste stream changed. It gave the Department of Health and Human Services (then known as the Department of Health, Education, and Welfare, or HEW) advisory authority over municipal waste programs and the funding of research and development. Policy makers believed that scientists could find technology to "fix" the garbage problem (such as the addition of toxic materials to household trash) by coming up with new methods of disposal. The Public Health Service was given jurisdiction over municipal waste, and the Bureau of Mines was tasked with the disposal of utility plant waste. The legislation also encouraged regional solid waste management planning and directed the secretary of HEW to provide 50 percent matching grants to states to survey their solid waste practices and to develop solid waste disposal plans. At this point, the federal government's primary goal was providing financial and technical assistance, not regulatory authority.

Air pollution concerns related to incinerators were addressed with the 1967 Air Quality Act, which required facilities to place pollution-control devices called scrubbers on the stacks that released smoke and fine particulate matter. The devices were expensive and led to the closure of many incinerator plants, even those that had benefited from new technology designed to reduce odors and smoke. Although incineration was considered a useful technology for dealing with waste, especially in urban areas, the concern over air quality was greater than its utility.

Three years later, the Resource Recovery Act was passed, reflecting the wave of public support for environmental protection that marked the 1970s. This time, Congress focused on recycling and hazardous waste—topics that had previously gone unrecognized as worthy of federal attention. The newly created EPA assumed responsibility for most issues through its Office of Solid Waste. The law moved beyond the 1965 statute by increasing the federal share of grants for planning and development of solid waste facilities from 50 to 75 percent. Funding was primarily targeted at refuse-derived fuel systems and demonstration projects. The EPA also began developing definitions and guidelines for handling specialized types of waste, such as hospital wastes that were difficult to segregate but were considered potentially contaminated. The statute also called for the EPA to study ways for implementing a system of national repositories for the storage and disposal of hazardous wastes, although the term *hazardous wastes* was not explicitly defined. The report, submitted to Congress in 1973, indicated that current efforts to manage hazardous waste were inadequate, with the EPA recommending that a national hazardous waste regulatory program be established (Davis 1993, 19).

Congress began its deliberations over the nation's waste policies in 1974, with hearings that emphasized the serious nature of the problems, especially the increase in the number of open dumps in the United States. But other issues, such as the oil crisis that started in 1973 with the embargo imposed by the Organization of the Petroleum Exporting Countries and energy production concerns, kept waste management from advancing further on the governmental agenda. With the resignation of President Richard Nixon and Democratic majorities seated in Congress, legislators returned to the business of environmental protection with two statutes that represented a change in policy direction. In 1975, Congress passed the Hazardous Materials Transportation Act, which was implemented by the federal Department of Transportation. The statute was enacted in part to address the illegal dumping of waste as many landfills refused to accept hazardous materials and as the habit of "midnight dumping" of waste became commonplace. Waste brokers found it relatively easy (and much cheaper) to just dump waste along roadsides or in rural areas rather than pay for disposal. Federal agencies were unprepared to handle both the amount and types of waste, but they were under pressure from environmental organizations to take a stronger role in dealing with disposal issues.

Responding once again to public opinion, Congress enacted the Resource Conservation and Recovery Act (RCRA) in 1976 as the environmental movement's advocacy for pollution control grew stronger. RCRA, an amendment to the 1965 SWDA, divided the management of waste into two categories, hazardous and nonhazardous, and directed the EPA to develop regulations to determine which facilities should be classified as sanitary landfills and which should be categorized as open dumps. The legislation was important because it provided a definition for household waste as "any solid waste [including garbage, trash, and sanitary waste in septic tanks] derived from households [including single and multiple residences, hotels and motels, bunkhouses, ranger stations, crew quarters, campgrounds, picnic grounds, and day-use recreation areas]." To the surprise of some, RCRA's definition of solid waste included liquids and contained gaseous material. Under this statute, most household waste could be disposed of in landfills, along with sewage sludge, municipal solid waste incinerator ash, agricultural wastes, and some mining wastes. Another important feature of RCRA was the development of a "cradle to grave" system that required management of wastes from the time they are created until they are safely disposed of, a concept that is an important element of hazardous waste management today.

The energy crisis of the early 1970s opened a window of opportunity for an improved form of incinerator called waste-to-energy (WTE), or mass-burn plants. By burning trash and other solid waste (even excess watermelons that could not be sold) at high temperatures, and excluding noncombustible materials such as appliances and dirt, water could be heated, turned into steam, and used to generate electricity. Because fossil fuels were in short supply and prohibitively expensive, engineers redeveloped the older type of combustion plants to produce power and reduce the amount of waste being sent to landfills. In order to provide an incentive for building new WTE facilities, Congress enacted the 1978 Public Utility Regulatory Policies Act. Local utilities were required to purchase a certain portion of their power from WTE, ensuring a market for the producers. The facilities were welcomed in areas of the country where landfills were rapidly filling up as the population density increased, such as California and the East Coast.

The deregulation policies of the Reagan administration, which began in 1981, led to a greater emphasis on privatizing solid waste management, rather than assigning the task to municipal government. Contractual services, provided by private companies, became more prevalent, especially in urban areas such as New York City. In 1980, approximately 60 of the new WTE incinerators were running and 200 were operational by 1985. But as quickly as the phenomenon and accompanying tax incentives had been created, the number of incinerators appeared to reach a plateau, with about 200 operational in 1990 (Walsh, Warland, and Smith 1997, 6–7).

As the price of oil dropped and public opposition to incineration developed, the issue of hazardous waste resurfaced. Hazardous waste issues (referred to as hazardous substances in the legislation) and radioactive wastes were addressed by Congress in 1980 with passage of the Comprehensive Environmental Response, Compensation, and Liability Act (CERCLA), also known as Superfund. The federal government was given the lead role in establishing a program to assess liability in cleaning up sites, and remediation was to be accomplished through creation of a $1.6 billion fund, which was expected to finance cleanup operations for five years. The monies were derived from a tax on chemical feedstock producers and from federal revenues. The EPA was also directed to develop the National Priorities List (NPL) of the most hazardous waste sites in the United States, which was to include at least one site from each state. The measure made its way through Congress and was signed by President Gerald Ford just before President Jimmy Carter took office.

The EPA attempted to meet a 1978 deadline for promulgating interim hazardous waste regulations, but the agency's failure to do so led to widespread criticism from Congress. Constituents concerned about sites in their neighborhoods put pressure on their elected representatives, who felt the EPA was devoting too few resources to hazardous waste programs. Despite these reservations, Congress reauthorized CERCLA in 1980. Its concerns were magnified throughout the early 1980s and were highlighted by a 1983 Office of Technology Assessment report indicating that problems were encountered in the collection of data about the amount and types of waste generated and the lack of incentives available to waste producers to consider pretreatment of recycling of wastes in lieu of land-based containment. A similar study by the National Academy of Sciences in 1983 pointed to a need for a clear definition of hazardous waste and training on waste reduction practices for companies generating hazardous waste (Davis 1993, 31–32).

The political momentum for policy change grew with the discovery of soil contaminated with dioxin from waste oil sprayed on roadways near Times Beach, Missouri, in late 1982. The EPA began taking samples of the soil and later warned residents about potential health effects from exposure. The problem became so serious that the community was evacuated, and eventually, the federal government purchased all of the homes and property in the area. While stories about hazardous waste dumping had appeared periodically in the media, they seemed to be isolated problems in the minds of most Americans. But Times Beach made headlines nationwide, focusing attention on the problem and again putting pressure on Congress to do something.

That "something" took the form of the 1986 Superfund Amendments and Reauthorization Act (SARA), which provided $8.5 billion through 1991 to clean up the most dangerous sites, set strict standards and timetables for their cleanup, and required companies to provide communities with information on the hazardous chemicals they used or emitted. Like the original Superfund legislation, funding for SARA came from a tax on petroleum and chemical feedstocks, a broad-based business tax, and federal revenues, plus cleanup expenses recovered from firms responsible for polluting a site on the NPL. The EPA was directed to work on 375 sites within five years, with an emphasis on some form of permanent cleanup that did not involve storing waste in landfills.

The public's concerns about energy and fossil fuels then began to eclipse concerns about solid waste management, and much of the EPA's interests shifted as well. Congress reauthorized the 1976 legislation in the form of the Hazardous and Solid Waste Amendments of 1984, expecting that technology-forcing methods of regulation (such as new standards for sanitary landfills) would lead to solutions that were more cost effective. The EPA, however, placed its faith in source reduction as a way of extending the life of landfills; this issue is discussed in more detail in Chapter 2.

A series of widely publicized events in the 1980s focused public attention on infectious and medical waste for a short period of time. The New York City Fire Department was responding to a warehouse fire in 1986, when officials discovered about 1,400 bags of medical waste that had been illegally dumped there. The company had submitted documents to the state's Department of Environmental Conservation stating that the wastes had been incinerated. A year later, a barge called the *Mobro 4000* left Islip,

New York, with more than 4,000 tons of garbage, headed for a landfill in North Carolina. Officials were tipped off that the barge carried medical waste, including hospital gowns, syringes, and diapers, and was looking for a place to unload. Once its mission was discovered, the barge headed back to sea, moving along the East Coast and finally traveling more than 6,000 miles before officials allowed it to dock to incinerate the waste. That same year, children playing in a trash bin in Indianapolis found vials of blood infected with the AIDS virus that had been improperly disposed of by a medical office. These events worried the public and led to the initial momentum for legislative action.

The most notable instances involving medical waste took place in 1988, when needles, syringes, and empty prescription bottles with New York addresses wound up on the shores of the Atlantic Ocean in New Jersey. Two months later, 10 miles of Long Island, New York, beaches were closed when medical waste washed ashore, and later that summer, similar incidents were reported along the coastlines from Maine to the Gulf of Mexico and along the Great Lakes. Congress responded with the Medical Waste Tracking Act, which mandated that the path of medical waste be followed for two years in participating states. The statute defined the term *medical waste;* established a "cradle-to-grave" tracking system; required standards for segregating, packaging, labeling, marking, and storing medical waste; and established record-keeping requirements and penalties that could be imposed for mismanagement.

RCRA was amended in 1996 with the Land Disposal Program Flexibility Act, which exempts hazardous waste from RCRA if it is treated to the point where it no longer exhibits the characteristics that make it hazardous. The flexibility component of the statute also applies to landfills that receive less than 20 tons of MSW and exempts small landfills in arid areas from groundwater monitoring requirements if no evidence exists of groundwater contamination.

Regulation of emissions from solid waste management facilities began in 1990 with passage of the Clean Air Act Amendments (CAA), requiring the EPA to develop rules to regulate sources, including landfills and hazardous waste incineration facilities, and the open burning of solid waste. The changes in the law were a result of studies showing that landfills and other facilities can be a major source of emissions that contribute to ambient ozone pollution, global warming, and concerns about toxic air pollution. Under the CAA, facilities must achieve the greatest degree of emission reduction possible through the application of best available control technologies and procedures. The EPA's regulations, finalized in 2000, apply to particulate matter, sulfur dioxide, hydrogen chloride, nitrogen oxides, carbon monoxide, lead, cadmium, mercury, dioxins, and dibenzofurans. Another set of regulations was issued in 1996 for existing municipal solid waste facilities, requiring the installation of landfill gas collection and control systems, which affected about 300 of the nation's largest landfill facilities. These regulations focused on methane gas, which also contributes to global warming and traps heat more effectively than carbon dioxide. Even though U.S. landfills are the smallest source of methane emissions created through the decomposition of organic refuse, the growth of developing countries is likely to increase landfill development. Perhaps more important, landfills are one of the few sources of methane gas that can be controlled, in comparison with those connected with domestic animals, rice production, and naturally occurring sources such as swamps and bogs (Tchobanoglous and Kreith 2002, 2.24–2.35).

As part of an integrated approach to waste management, the federal government has also attempted to encourage voluntary source reduction, mostly through financial support of pilot programs. In 1996, Congress enacted the Bill Emerson Good Samaritan Food Donation Act, which protects businesses, organizations, and individuals who donate food in good faith from any legal liability that may arise. Donations to food banks from restaurants, for instance, may not only feed hungry people but also reduce the mass of organic material that would otherwise be dumped into a landfill. Most source-reduction initiatives are regulated by the EPA, and some states and local governments provide technical or financial support, such as grants.

Since 1965, the states have played a secondary role to the federal government in waste management, primarily because they have been the recipients of funding (grants) and have been reactive in responding to federal regulations. Programs such as matching grants for planning and demonstration projects made states reliant on federal direction, and it was not until the passage of RCRA in 1976 (reinforced by the 1984 RCRA amendments) that the government provided legislative guidance on the preparation of state solid waste management plans. The EPA was delegated the authority for approving state plans according to six requirements for compliance, including the identification of authorities responsible for plan implementation, the prohibition of new dumps, and the closing and upgrading of existing dumps.

Environmental Justice and Waste Management

Opposition to incineration was loosely organized in the early 1980s, driven by residents who did not want a facility in their neighborhood, an issue that would also be faced by municipal governments seeking sites for new landfills. Concerns were raised about the dangers posed by toxic air emissions when trash was burned and over the issue of what to do about incinerator ash. Critics argued that incinerator operators could not effectively separate hazardous waste from MSW and that carcinogens such as furans and dioxins were being released into the air. A debate erupted over whether the ash produced from burning processes should be classified as hazardous and sent to specially designed landfills, rather than mixed in with other waste, which would significantly increase the cost of incineration.

In 1984, a 90-page study, the Cerrell Report, was written for the California Waste Management Board, identifying ways to overcome the political obstacles to siting waste-to-energy facilities. The report suggested that facility proponents look at community demographics to determine which areas would be the least resistant to projects, noting that facilities would be likely to be accepted in rural areas with politically conservative residents who were above middle age, had an education level of high school or less, and would receive significant economic benefits from the facility (Walsh, Warland, and Smith 1997, 13–14).

When a copy of the Cerrell Report was obtained by anti-incineration forces, opposition to various types of waste management treatment really began to coalesce. The industry acronym LULU (locally unwanted land uses) was used to describe facilities that no one wanted; the acronym describing these residents' sentiments is known as NIMBY (not in my backyard). Opposition developed throughout the United States in the mid-1980s, built around what has been referred to as an "equity movement"—a gradual mobilization around long-standing grievances among members of a culturally identifiable collectivity, such as African

Americans, Native Americans, or women—which depends for its success on the appropriate combination of political opportunities and internal organization among those involved (Walsh, Warland, and Smith 1997, 45).

Sociologist Robert Bullard, who has authored numerous books on the issue of environmental equity, or environmental justice, as the concept is more commonly known, identified this issue after studying the spatial location of all MSW facilities in Houston as part of a class-action lawsuit. The litigation stemmed from an attempt to site a municipal landfill in a suburban, middle-class neighborhood of single-family homeowners. The neighborhood, Northwood Manor, was made up of more than 82 percent African American families, and Bullard determined that it was not randomly selected for the landfill, as was the case throughout the South. He found that garbage dumps, hazardous waste landfills, incinerator operations, smelters, paper mills, chemical plants, and other polluting facilities were being imposed on specific types of neighborhoods: "These industries have generally followed the path of least resistance, which has been to locate in economically poor and politically powerless African American communities" (Bullard 2000, xiv–xv).

The first national protest on the issue of hazardous waste facility siting took place in 1982, when black residents in Warren County, North Carolina, united in opposition to a proposed dump for 32,000 cubic yards of soil contaminated with highly toxic polychlorinated biphenyls (PCBs). The soil had been illegally dumped along roadsides by a waste hauler in 1978, collectively constituting the largest spill ever documented in the United States. The decision to bury the soil in Warren County was made by the state's governor, James Hunt, and the site was later referred to by residents as "Hunt's Dump." Warren County had the highest percentage of black residents in the state and had one of the lowest levels of median family income. Local grassroots groups joined with national organizations to oppose the facility. During protests, hundreds of protesters were arrested, marking "the first time anyone in the United States had been jailed trying to halt a toxic waste landfill" (Bullard 2000, 31).

Environmental justice issues involving waste management gained credibility with two critical reports issued in the 1980s. The first was a 1983 report by the U.S. General Accounting Office (now called the Government Accountability Office) that found a strong relationship between the siting of four hazardous waste landfills and the race and socioeconomic status of surrounding communities in Alabama, South Carolina, and North Carolina. The second study was a 1987 publication of the United Church of Christ's Commission for Racial Justice titled *Toxic Wastes and Race in the United States: A National Report on the Racial and Socioeconomic Characteristics of Communities with Hazardous Waste Sites.* It further documented the relationship between race and siting, calling upon organizations to unite against the problem of environmental racism.

Numerous examples exist of how environmental justice and waste management issues intersect. For instance, in 1990, Browning-Ferris Industries (BFI) began siting studies for a garbage transfer and recycling facility in the predominantly black neighborhood of Titusville in Birmingham, Alabama. The company wanted to build on vacant property owned by a snack foods company, which operated on an adjacent lot. Although BFI followed the letter of the law in obtaining the necessary permits for the facility, the nearby residents were never informed of the nature of the operation. Once they learned about the trash facility, opponents raised concerns about drinking water quality and the umbrella of secrecy under which the plant had quickly been built. In 1994, activists marched on Birmingham's City Hall to protest the siting of the BFI plant, resulting in a confrontation between police and protesters. Although the company was later found to have acted improperly, the city refused to stop the plant from operating (Westra 2001).

Bullard (2000) notes that the mission of the EPA never included addressing issues of environmental policies and practices that result in unfair, unjust, and inequitable outcomes. The federal government took action in the 1990s, he says, only because of mounting scientific evidence and prodding by grassroots organizations and educators. The federal government's role became solidified only when President Bill Clinton signed Executive Order 12898, "Federal Actions to Address Environmental Justice in Minority Populations and Low-Income Populations," in 1994. The measure reinforced sections of the 1964 Civil Rights Act and encompassed the 1970 National Environmental Policy Act by requiring government agencies to consider environmental justice concerns in assessing the impact of their projects and programs. The EPA went a step further in 1997 by establishing the National Advisory Council on Environmental Policy and Technology to examine facility permitting.

Since that time, environmental justice issues have arisen occasionally, but they have been a part of overall concerns about how to manage waste, especially when sites are selected in areas with a large minority population, a topic that is addressed in more detail in Chapter 2.

References

Bryant, Chris. 2007. "Fuels from the Forests." *Forest Magazine* (Spring): 32–37.

Bullard, Robert D. 2000. *Dumping in Dixie: Race, Class, and Environmental Quality.* Boulder, CO: Westview.

Bureau of International Recycling. 2007. "About Recycling." [Online article; retrieved 4/16/07.] www.bir.org/aboutrecycling/.

Colten, Craig E., and Peter N. Skinner. 1996. *The Road to Love Canal: Managing Industrial Waste before EPA.* Austin: University of Texas Press.

Davis, Charles. 1993. *The Politics of Hazardous Waste.* Englewood Cliffs, NJ: Prentice-Hall.

Forest Preserve District of Cook County, Illinois. 1971. "Paper Bags." Nature Bulletin No. 438-A. [Online article; retrieved 4/29/07.] www.newton.dep.anl.gov.

Great Idea Finder. "Shopping Bag." [Online article; retrieved 4/29/07.] http://www.ideafinder.com/history/inventions/shopbag.htm.

Groff, Rod. 2007. "A Greener Way to Hydrate." *Arizona Republic,* July 7, B-3.

Hook, Paula, and Joe E. Heimlich. 2007. "A History of Packaging." Ohio State University Fact Sheet. [Online information; retrieved 4/29/08.] http://ohioline.osu.edu/cd-fact/0133.html.

Luton, Larry. 1996. *The Politics of Garbage.* Pittsburgh: University of Pittsburgh Press.

McCutcheon, Chuck. 2002. *Nuclear Reactions: The Politics of Opening a Radioactive Waste Disposal Site.* Albuquerque: University of New Mexico Press.

Melosi, Martin. 2001. *Effluent America: Cities, Industry, Energy and the Environment.* Pittsburgh: University of Pittsburgh Press.

Rogers, Heather. 2005. *Gone Tomorrow: The Hidden Life of Garbage.* New York: New Press.

Tchobanoglous, George, and Frank Kreith. 2002. *Handbook of Solid Waste Management,* 2nd ed. New York: McGraw-Hill.

Turnberg, Wayne L. 1996. *Biohazardous Waste: Risk Assessment, Policy, and Management.* New York: John Wiley & Sons.

U.S. Environmental Protection Agency (EPA). 1982. *Guide for Infectious Waste.* Washington, DC: EPA.

U.S. Environmental Protection Agency (EPA). 2007a. "Basic Information: Municipal Solid Waste." [Online information; retrieved 4/17/07.] http://www.epa.gov/msw/facts/htm.

U.S. Environmental Protection Agency (EPA). 2007b. "Solid Waste Landfills." [Online information; retrieved 4/17/07.] http://www.epa.gov/garbage/landfill/sw_landfill.htm.

U.S. Environmental Protection Agency (EPA). 2007c. "Using Phytoremediation to Clean Up Sites." [Online information; retrieved 7/5/07.] http://www.epa.gov/superfund/accomp/news/phyto.htm.

Walsh, Edward J., Rex Warland, and D. Clayton Smith. 1997. *Don't Burn It Here: Grassroots Challenges to Trash Incinerators.* University Park: Pennsylvania State University Press.

Westra, Laura. 2001. "The Faces of Environmental Racism: Titusville, Alabama and BFI." In *Faces of Environmental Racism: Confronting Issues of Global Justice,* edited by Laura Westra and Bill E. Lawson. Lanham, MD: Rowman and Littlefield.

World Health Organization (WHO). 1999. *Safe Management of Wastes from Health-Care Activities.* Geneva: World Health Organization.

2

Problems, Controversies, and Solutions

Paper or Plastic?

The question "Paper or plastic?" is illustrative of the problem that many consumers face as they deal with waste management issues. As they approach the checkout counter of the typical grocery store, they realize that they must make a decision as to which kind of bag they will choose to contain their purchases. A customer may consider whether or not a bag is made from a sustainable source, or whether or not it is biodegradable, or the ease of handling, or reuse, or any one of more than a dozen issues that are both environmentally and politically sensitive. So which is the "greener" choice for consumers?

Producers of both types of products argue that the facts are on their side and that their product is better for the environment than the other. A plastic bag begins its ubiquitous life as a form of fossil fuel, made from crude oil, natural gas, or some other petroleum derivative. Molecules are formed into polymers that are heated, cooled, and then flattened to just a few millimeters in thickness. Trillions of plastic bags are made every year; about one-quarter of all plastic bags are produced in Asia, while North America and Europe account for nearly 80 percent of their use. Consumers use them for hauling away trash and garbage, for carrying groceries and produce, for carrying retail items from the store to the car, and for other routine purposes. The Worldwatch Institute estimates that Americans throw away some 100 billion polyethylene plastic bags each year, with less than 1 percent of them recycled (Worldwatch Institute 2007).

One reusable bag company estimates that the average family accumulates 60 plastic bags in only four trips to the grocery store. The bags are lightweight, do not tear as easily as paper bags, are distributed freely throughout the produce section of stores, and can be reused for purposes ranging from an impromptu ice pack to making water balloons. Most of them end up being thrown away, although that does not mean that they become part of the waste stream. They are flimsy and easily held aloft by winds that can carry them for miles across the landscape, where they catch on trees and other vegetation; get trapped along fence lines; become tucked into niches in rocks; clog sewers, drainage ditches, and waterways; or worse yet, are ingested by wildlife. Plastic bags do not biodegrade—they photodegrade when they are exposed to sunlight, although no one is quite sure how long that may take, as plastic bags have been a part of the waste stream for less than 50 years. To estimate how long it would actually take, researchers use respirometry tests. They place a sample of a substance, such as a banana peel, in a vessel containing microbe-rich compost, and then aerate the mixture. Over the course of several days, the microorganisms assimilate the sample bit by bit and produce carbon dioxide as an indicator of degradation.

While the process works for organic matter such as banana peels (which take several days to break down), nothing happens when the tests are run on plastic bags. No carbon dioxide is produced and no degradation takes place because polyethylene is a human-made polymer that microorganisms do not recognize as food. But when plastic bags are exposed to ultraviolet radiation, as they would be when exposed to sunlight, the polyethylene's polymer chains become brittle and start to crack. This result suggests that they eventually will break down into microscopic pieces, but no precise estimates are yet available on how long that process will take (Lapidos 2007).

A new plastic bag–related problem has developed in the subtropical oceans of the world, where currents move in a clockwise spiral, or gyre, which traps debris that originates on land. The five major subtropical gyres on Earth are created by airflows from the tropics toward the polar regions. In these deep ocean regions, trash and debris accumulate and are trapped in the maelstrom and cannot be broken down by bacteria or other marine microorganisms. One area of floating plastic debris is estimated to be as large as the size of Texas, with objects in the gyre circulating for 16 years or more. A single ton of debris collected by researchers included a drum of hazardous materials, an inflated volleyball half covered in gooseneck barnacles, a plastic coat hanger, a cathode-ray tube for a television, an inflated truck tire, polypropylene net lines, and pieces of plastic and glass. Plastic resin pellets collected indicate that they accumulate toxic chemicals such as DDT (dichloro-diphenyl-trichloroethane), an insecticide banned in the United States; the pellets are ingested by jellyfish, which are then eaten by fish, and so on up the food chain. One study estimates that the plastic polymers commonly used in consumer products are indigestible by organisms, and ecologists believe it may take as long as 500 years for degradation by sunlight to offset the slow oxidation of their constituent parts (Moore 2003).

Other types of packaging containers, such as paper bags, have been a part of society for a much longer period of time. The paper industry of the late 1600s and early 1700s relied on fiber from recycled cotton and linen, but rag shortages at that time led to a search for a different source and ways to make paper. The answer came in the mid-1800s with the discovery of a reliable way to make paper from tree pulp. The process allowed manufacturers to make bags that were relatively inexpensive, stored easily, and could be made wherever a source of wood was readily available.

Because trees are considered a sustainable natural resource, paper is perceived as a more environmentally sound way of dealing with consumer demand for bags. Paper biodegrades even when placed in landfills, unlike plastic. The same type of respirometry tests used on banana peels have an effect on newspapers and other paper products (newspapers take two to five months to biodegrade in a compost heap, for instance). Paper can be de-inked and reprocessed into pulp, which can then be made into another paper product, although the life cycle is not infinite. Consumers routinely recycle paper products, including paper bags, but especially newspapers, so paper bags are less likely to end up in landfills or on the landscape. They are less harmful if swallowed or eaten by wildlife and are less likely to clog drainage ditches or other waterways because the fibers in paper break down in water. But paper bags take more energy to produce, making them more expensive than plastic bags, and are being phased out of use by some retailers, who have opted for the lighter weight (and less expensive) plastic bags.

So Which Bag Do You Choose?

Studies that look at the life cycle energy analyses of paper versus plastic consider several factors, such as the total energy used to manufacture the bag and the energy of its physical materials (transportation, electricity, fuel extraction, and processing), the amount of pollutants produced (airborne and waterborne), carrying capacity, recycling rates, and bagging techniques. If the primary consideration is solid waste, bigger definitely is not better. Compared with paper grocery bags, plastic grocery bags (Film and Bag Federation 2007):

- consume 40 percent less energy than paper
- generate 80 percent less solid waste
- produce 70 percent fewer atmospheric emissions
- release up to 94 percent fewer waterborne wastes

Some retailers have attempted to deal with the issue by developing and selling alternatives so that consumers will not have to make the hard decisions on which bag to choose. In July 2007, Whole Foods markets in the New York area started selling $15 cotton bags designed by Anya Hindmarch, who was previously better known for her $1,500 purses. The company ordered 20,000 of the Hindmarch bags and had to limit sales to three to a customer while supplies lasted because of demands by savvy shoppers. In Taiwan, sales of her "I'm not a plastic bag" bags caused a stampede of would-be purchasers that sent 30 people to the hospital and brought out riot police (Burros 2007).

While the paper versus plastic debate is ongoing, it is just one facet of a growing international problem. After centuries of dealing with waste, humanity could be expected to have developed numerous successful strategies that would eliminate the problems waste causes, or eliminate waste entirely. However, that is not the case. As the population grows, so, too, does the amount of waste created, especially waste packaging like paper and plastic. The next section identifies some of the key issues being addressed by governments, trade associations, and nongovernmental organizations and discusses how well new strategies are working. The chapter's discussion is limited to issues that face the United States; the international perspective is covered in Chapter 3.

Reducing the Amount of Waste

The amount of trash being created can be reduced in many ways, but waste management experts believe that the first way to address the mountains of waste being created is through source reduction, sometimes called pollution prevention, described in detail in Chapter 1. Source reduction, defined as a net reduction in the generation of waste compared with the previous amount of waste generated, is almost always at the top of the waste management hierarchy. Strategies include developing products that are durable, reusable, and remanufactured; products with no, or reduced, toxic constituents; and products marketed with no, or reduced, packaging.

The last option is that which many consumers recognize as source reduction. One of the early examples of this strategy is the transition that occurred after record companies began releasing music compact discs (CDs) in bulky cardboard packaging called longboxes. These containers were designed to fit into the same size retail slots as vinyl record albums, and took up a similar amount of space. The longbox allowed music producers to use the same approach to artwork and graphics featured on record album covers, along with liner notes that included song lyrics, information about the artist and musicians, and marketing material. The longbox was often twice the size of the product itself, and although the extra packaging was unnecessary, it did give customers the perception that they were getting more for their money.

The CD packaging that consumers purchased was usually thrown away somewhere between the cash register and the car stereo with virtually no possibility of being reused. Pressure by environmental groups during the 1990s resulted in the longbox being abandoned by the end of the decade in favor of the same hinged polystyrene jewel cases that had been introduced in the 1980s. A single media tray contains the disc, and the hinged lid makes for easy access. The two-part design of the jewel box keeps the disc from being scratched yet allows visibility for making a music selection. A slimline version of the jewel case is also in use now, as are digipacks, a generic term used for cardboard CD cases that are less durable but considered more environmentally sound, and CD sleeves similar to envelopes.

Compact discs (and their relatives, such as DVDs) illustrate the way source reduction often works. Changes can be made in both industrial production methods and in packaging, and sometimes a product can be totally eliminated. Since 1977, for instance, the weight of two-liter plastic soft drink bottles has been reduced from 68 grams each to 51 grams, resulting in a reduction of 250 million pounds of plastic per year in the waste stream (U.S. EPA 2007b). In Alameda County, California, the General Service Agency eliminated the cost of printing new letterhead by switching to a word processing program that uses templates. Employees type their letters and memos on the appropriate computer template, which can be customized for every office.

Another source reduction strategy is to look for ways to use different types of materials from the perspective of ownership cost, or a product's entire life cycle. The city of Berkeley, California, bought single-polymer plastic lumber benches for city streets and parks because of their durability. They are made of composites that make it easier to clean off graffiti and to repair holes and damage. Initially, the benches cost the city more than it would have paid for wood benches, but they are much less expensive to maintain because they never need to be sanded or painted. Similarly, purchasing rubber playground surfaces is initially more expensive than using sand or other loose fill material, but the rubber equipment eliminates the need to continually clean and replace the loose fill. Rubber surfaces have the added benefit of making playgrounds more accessible for children with disabilities (Conservatree 2007).

In a modified version of source reduction, California has adopted a "zero waste" philosophy that is based on the concept that wasting resources is inefficient, and efforts should focus on environmental stewardship and redefining the concept of waste in society. As the California Integrated Waste Management Board describes it:

> California is a state rich in natural resources and has an environment unlike any other, and those resources need to be protected. In that effort, zero waste California stretches beyond our previously imagined goals. It is the ultimate in environmental stewardship—and a goal we can all work together to accomplish.

> Californians know how to "reduce, reuse, and recycle." We have been living it and have come to make it part of our everyday lives. Now, with recycling and conservation programs in every city, we are able to embrace the zero waste concept as our guiding principal and goal for the future. (California Integrated Waste Management Board 2006b)

Enhancing and Improving Recycling

Recycling initially became popular in the 1960s as a way to reduce the amount of trash going to landfills and incinerators just as the environmental movement was gaining momentum. It was also driven by the grassroots, anti-incineration pressures of the late 1980s, indicative of the ebb and flow pattern of support for recycling over the last 40 years. Curbside recycling was established in many communities during the 1980s as residents started using different colored receptacles for paper, plastic, and glass recyclables. By the 1990s, industries jumped on the recycling bandwagon, primarily because they could promote their products and their green image at the same time.

In 2005, the U.S. Environmental Protection Agency (EPA) reported that nearly 40 percent of containers and packaging were recycled, led by steel packaging (mostly cans) at 63 percent, paper and paperboard at 50 percent, and aluminum at 36 percent, with 45 percent of all aluminum beverage cans recycled. Approximately 25 percent of glass containers; 15 percent of wood packaging, mostly pallets; and 9 percent of plastic containers and packaging, mostly from soft drink, milk, and water bottles, were recycled. The most recycled materials were paper products, such as newspapers (89 percent), high-grade office papers (63 percent), and magazines (39 percent). Unwanted "junk" mail was recovered at a rate of 36 percent and 18 percent of telephone directories were recycled (U.S. EPA 2007a).

But the fact that material is collected for recycling does not necessarily mean that it is reprocessed, one analyst notes. "Almost half of discarded newspapers and office paper is buried or burned, while two-thirds of glass containers and plastic soda and mile bottles are trashed instead of recycled" (Rogers 2005, 176–177). Reprocessing can be expensive, and it can also become less feasible the more often the material goes through the process. Plastic, for instance, is the least recyclable of materials because it loses its flexibility when made molten again. Paper fibers begin to get shorter with each successive use, making them less able to hold together. The U.S. Food and Drug Administration prohibits the use of old food packaging to make new food containers; studies show that even when heated at very high temperatures, it is impossible to sterilize the packaging. A second problem deals with contamination; one type of plastic can contaminate an entire truckload of another type, such as mixing soda bottles with mixed plastics, causing new bottles to be yellowed or to have black streaks. In that sense, technical issues place limits on what can be done with old newspapers or plastic water bottles that actually have a finite life span.

> Even if recycled under the best of conditions, a plastic bottle or margarine tub will probably have only one additional life. Since it can't be made into another food container, your Snapple bottle will become a "durable good" such as carpet or fiberfill for a jacket. Your milk bottle will become a plastic toy or the outer casing of a cell phone. Those things, in turn, will eventually be thrown away. "With plastics recycling, we're just extending the life of a material." (Gurnon 2003)

In addition, not all recyclables stay where they are collected. It is estimated that at least 20 to 30 percent of plastic recyclables leave the United States for Asia, a journey that requires additional energy for transport and material management. An analysis of what happens to plastics in Humboldt County, California, shows that consumers may not fully understand what happens when they toss their plastic soda bottle into the recycling bin. The items collected at two sites there are routinely baled and stored for about a month until they collectively make up a 12-ton truckload. The bales are transported to a facility in Sacramento, California, where they are sold to brokers in Hong Kong, who pay to transport them by container ship from the Port of Oakland (another truck trip) to China. Once in China, the bales go to recyclers in Shanghai and Guangdong Province, the closest province to Hong Kong. The facilities where the bales are processed range from "mom and pop" operations to large, comparatively modern facilities. Worker safety standards are virtually nonexistent in that region of China, according to the Basel Action Network, cited in Gurnon, and materials are processed in workplace conditions that are often dirty and primitive (2003).

Despite the fact that household recycling is a relatively simple procedure, the practice has not been universally accepted by the public. But although some operational problems occur in expanding consumer recycling, as explained previously, studies indicate that programs can be successful in those areas where the practice is coupled with public education. Once consumers become aware of how to recycle used goods, they also need to understand the benefits of changing their behavior. Sometimes, the benefit might be a cash incentive (such as collecting aluminum cans with a bounty priced per pound) or environmental protection (not mixing hazardous waste with household trash so that no additional environmental damage occurs). Consumers must also have confidence in the quality of products made from recycled materials and be prepared to pay extra for them, as they are usually more expensive than their counterparts made from virgin materials.

Some items that consumers routinely throw into recycling bins cause problems when attempts are made to reprocess them. An example is pressure-sensitive adhesives (PSAs) or self-adhesives such as stamps, sticky notes, and labels. They do not require moisture to activate them (just peel and stick). But most paper recycling systems use water as a medium to transform recovered paper back into pulp, which is then reprocessed into paper. The PSAs do not dissolve in water and deform under heat and pressure, making it difficult to screen them out from the pulp. They can become lodged on the papermaking equipment, causing serious mechanical damage, and can create holes or weak spots in paper that can jam copiers and printers. As is the case with mixed bales of plastic recyclables, paper can be contaminated by an excessive amount of PSAs (California Integrated Waste Management Board 2006a).

To solve this "sticky" problem, the U.S. Postal Service is working with researchers to develop a universally recyclable PSA. The Postal Service, itself a major supplier of PSA products (accounting for about 14 percent of U.S. consumption, mostly postage stamps) notes that consumers prefer the moisture-free versions of stamps, so its operations, too, would need to change in order to reduce the amount of PSAs in the waste stream. Indeed, consumers can use numerous alternatives to PSAs, such as printing addresses directly on labels, hand printing addresses on large mailing envelopes, avoiding the use of sticky notes, such as Post-it brand notes, and requesting and purchasing moisture-activated postage stamps (California Integrated Waste Management Board 2006a).

Recycling participation rates vary considerably from one neighborhood to another, based on factors such as whether trash and recyclables are collected on a weekly or monthly basis, whether trash and recyclables are picked up at the same time or on the same day, if the recycling program is voluntary or mandatory, and whether residents pay extra for borrowing a recyclable bin for collection. In a truly sustainable recycling economy, for each truckload of recyclable commodities leaving a region, a truckload of recycled goods must enter (Tchobanoglous and Kreith 2002). When consumers must pay extra for the loan of a recycling container or bins (typically about $5 more per month), any incentive to recycle may be lost in simple pocketbook economics. It becomes cheaper, and thus easier, simply to toss recyclables in with the rest of the household's trash.

Multifamily dwelling units pose a special problem for recycling advocates, especially as housing costs increase and purchasing a home becomes less feasible for many people. The United States has about 100 million occupied residential units, about 16 million of which are located in buildings or complexes with five or more units. Residents in these buildings may be left out of curbside recycling programs because the nature of their homes may not allow them to leave their recyclables at the curb for collection. In some apartment complexes, for example, building layouts require residents to take their trash to a centralized collection shoot or trash dumpster rather than separating it out (De Young, 1995).

Another aspect of recycling involves incorporating materials that are not generally recycled by consumers, and many of the efforts involve partnerships between the government and industry. For example, the Carpet America Recovery Effort (CARE) was established by members of the carpet industry, representatives of government agencies, and nongovernmental organizations in January 2002. It is an attempt to reduce the amount of old carpet going to landfills and to increase the reuse of postconsumer carpeting. The organization monitors the progress made toward national goals for carpet recovery and provides technical assistance in recovering value from discarded carpet. The voluntary agreement involves enhancing ways of collecting postconsumer carpet and developing market opportunities for recovered carpet. The agreement, which took two years to negotiate, is considered a model for future product stewardship initiatives (CARE 2007b).

Carpet represents a unique challenge for recycling and reuse. In 2002, an estimated 4.7 billion pounds of waste carpet was discarded, with about 96 percent going to landfills. This becomes a problem as landfill disposal capacity shrinks, especially considering carpet's bulk, which also makes it difficult and expensive to handle. These are just two reasons why the industry has teamed with government agencies to find ways to divert carpet from landfills or other means of disposal, such as waste-to-energy (WTE) facilities. The group's goal is to increase the amount of carpet that is recycled from 3.8 percent in 2002 to 20 to 25 percent in 2012, and to increase the number of pounds of carpet that are reused from 0 in 2002 to 203 million to 339 million pounds in 2012 (Minnesota Office of Environmental Assistance 2007).

According to CARE's May 2007 *Annual Report,* 261 million pounds of postconsumer carpet were diverted from landfills in 2006, a 16 percent increase over diversion in 2005, and 240 million pounds were recycled, a 23 percent increase over 2005. But the effort resulted in less diversion and recycling than the organization expected for this phase of the recovery effort. Two factors appear to be the causes of the missed goals: a dramatic drop in sales in the residential carpet sector due to the drop in new housing starts and a reduced responsiveness in the results used to generate the survey (CARE 2007a).

Dealing with Beverage Cans and Bottles

Source reduction and recycling take many forms, but one of the easiest ways consumers can participate in reducing waste is by cutting back on the number of plastic beverage bottles and drink cans that end up in landfills. According to the Container Recycling Institute, Americans spent more than $270 billion on fountain and packaged beverages in 2005, about the same amount they spent on gasoline that year. In simple terms, the number of packaged beverages sold increased by one extra bottle or can each week for every man, woman, and child in the United States (Gitlitz and Franklin 2007).

Prior to the 20th century, beverages like beer were served from draught (on tap) in bars and restaurants. Producers then began selling beer in glass bottles, which could be purchased, brought back to the seller, and then refilled. Local and regional breweries dominated the marketplace, shipping their beer in kegs and refillable containers. The development of containers for beverages was revolutionized in 1930 with the introduction of the first one-way, throwaway can and the popularity of packaged beer. During World War II, U.S. companies shipped millions of cans of beer to military personnel stationed overseas. One-way glass bottles made their appearance on the market in the 1940s; in 1947, refillable glass bottles had a 100 percent market share, but by 2000, the market share was less than 1 percent. The Air Force used one-way bottles to ship beer to the military toward the end of World War II, creating a new class of consumers: veterans. Canned soft drinks could be found on grocery store shelves in the 1950s, but they were primarily private-label soft drinks produced in steel cans that some consumers complained had a metallic aftertaste. In 1957, the aluminum cans began to replace steel containers; the aluminum can had actually been invented during the Great Depression, but consumers did not accept an item designed to be thrown away during a period when all resources were considered valuable. From 1959 to 1969, Coca-Cola and Pepsi promoted one-way glass bottles for their products and instilled in consumers a habit of discarding beverage containers, resulting in declining return rates for refillable containers. The replacement of steel cans with aluminum cans and the introduction of plastic bottles in the 1970s further diminished the popularity of the refillable glass bottle for beer and soft drinks. "The one way container not only liberated consumers from returning bottles, but also liberated retailers from the burden of managing deposit-return systems and bottlers from having to wash and inspect returned bottles" (GrassRoots Recycling Network 2002).

Similarly, milk was historically distributed to consumers in glass bottles, which were delivered to each home, picked up again when empty, returned to the dairy, and then refilled. In 1906, milk suppliers began delivering their product in one-way paper cartons instead of glass bottles, with plastic-coated cartons appearing in the mid-1930s. The one-way plastic jug debuted in 1964, made of high-density polyethylene (HDPE). Another type of plastic used for refillable milk containers, polycarbonate, appeared in the late 1970s, and during the 1980s and 1990s, U.S. dairies provided milk in these types of containers to schools and other large institutions. When home delivery ended and consumers began buying their milk at grocery stores, one-way containers were used only to sell private-label milk, usually from local dairies.

Since that time, one-way containers have become the norm, with refillables the exception. Massachusetts leads the rest of the nation in the use of refillable bottles, with 16 percent of the beer sold in the state. Home delivery of milk in refillable glass bottles is still available in some portions of the United States, and in some retail chains and natural foods markets, refillable milk bottles are available for purchase. However, the beverage market is changing dramatically, and bottled water sales are among the highest in growth. These bottles (now mostly plastic) result in mounting litter and waste problems across the country.

To understand the magnitude of this problem, consider the follow statistics. Nonsparkling bottled water sales doubled in three years, from 15 billion units sold in 2002 to 29.8 billion sold in 2005, which is almost seven times the 3.8 billion units sold in 1997. Sales of plastic water bottles one liter or less increased more than 115 percent, from 13 billion in 2002 to 27.9 billion in 2005. Although soft drinks are a critical part of the beverage sector's overall sales, bottled water and other nonalcoholic, noncarbonated drink sales (such as tea) are likely to surpass soda by the end of this decade (Gitlitz and Franklin 2007).

The scale of beverage waste (bottles and cans that are not recycled and end up in landfills and incinerators) is staggering. In 2005, an estimated 146 billion containers ended up as litter alongside streets and highways, on beaches and riverbanks, in parks and fields, and in vacant lots. This figure includes approximately 54 billion aluminum cans; 52 billion plastic bottles and jugs; 30 billion glass bottles; and 10 billion pouches, cartons, and drink boxes (Gitlitz and Franklin 2007).

One contentious policy option has been the shift from refillable bottles to container deposit legislation. The first deposit law in the United States was enacted in 1934 by the National Recovery Administration, which required deposits of two cents for small bottles and five cents for large ones after bottlers had been using the noncollection of deposits as a competitive weapon. The public's concerns about litter during the 1950s led to the first statewide restriction on beverage containers, Vermont's ban on nonrefillable beer bottles, which was enacted in 1953. But because the public perceived that the ban did not, in fact, reduce littering, the ban was not renewed in 1957 and was allowed to lapse. By the late 1960s and early 1970s, when the environmental movement became most active, the issue reemerged because one-way containers made up 60 percent of the volume of soda pop containers sold in the United States and about 75 percent of those for beer. More than 350 bans or taxes on nonrefillable containers were introduced at the federal, state, and local levels (GrassRoots Recycling Network 2002). The antilittering campaigns were led by Lady Bird Johnson, first lady and wife of Lyndon B. Johnson, who took over the presidency after the assassination of John F. Kennedy in 1963. She was considered one of the earliest advocates of highway beautification after she mounted a nationwide campaign against littering.

Oregon enacted a bottle bill in 1972, and during the 1970s, four other states did so. In the 1980s, the antiregulation mood of the country seemed to dampen consumer enthusiasm for mandatory beverage deposit legislation. Thus far, 11 states have implemented bottle bills that provide a cash rebate for deposits consumers pay at purchase when they return and recycle cans and bottles: California, Connecticut, Delaware, Hawaii, Iowa, Maine, Massachusetts, Michigan, New York, Oregon, and Vermont. In 2007, campaigns to enact bottle bills were underway in Arkansas, Illinois, Maryland, North Carolina, South Carolina, Tennessee, and West Virginia. Three states were considering legislation to expand their existing legislation: Connecticut, Massachusetts, and New York. Only one bottle bill has ever been repealed: that enacted by the city of Columbia, Missouri (Container Recycling Institute 2007).

With the exception of Delaware, which exempts aluminum cans, the states with bottle bills require refundable deposits, ranging from 1.5 cents in California to 10 cents in Michigan, on all beer and carbonated soft drink cans and bottles. In 8 of the 11 states, deposits cover beer and soda only; California, Hawaii, and Maine also cover carbonated juices, iced teas, sports drinks, and bottled water. In five states (California, Hawaii, Massachusetts, Michigan, and Maine) some or all unclaimed deposits become the property of the state and are used for recycling education, to fund environmental programs, to administer the deposit system, or as a supplement to the state's general fund. Bottlers and distributors in the other states keep the unredeemed deposits. Research indicates that container deposit laws do have an impact on recycling rates. In states with bottle bills, an average of 490 containers per capita are recycled, compared with only 190 containers per capita in the remaining 39 states, or 150 percent fewer containers (Center for Policy Alternatives 2007).

Efforts to increase beverage container recovery have expanded in the 21st century, due in part to concerns raised by cooperative coalitions such as the group Businesses and Environmentalists Allied for Recycling (BEAR). The group warns that in the absence of new recovery and market development initiatives, beverage container recycling rates are likely to steadily decline in future years. In 1999, an estimated 193 billion aluminum, polyethylene terephthalate, HDPE, and glass beverage containers were generated in the United States, equivalent to 11.1 million tons, or 684 containers per capita. Of this amount, 78.1 billion containers were recycled (3.4 million tons, or 277 containers per capita) and 114.4 billion containers were disposed of (7.7 million tons, or 407 containers per capita).

A 2002 study by BEAR notes that significant environmental benefits can be realized by recycling beverage containers. Energy use is reduced, as are greenhouse gas emissions. The energy saved is equivalent to more than 32 million barrels of oil per year, dominated by the benefits associated with recycled aluminum cans. The use of land for disposal of waste and for the extraction of virgin materials is reduced, and litter is reduced, leading to reduced human injuries and avoided harm to farm machinery and animals. In addition, the development and operation of a beverage container recycling infrastructure creates a significant number of jobs and has been shown to improve U.S. competitiveness (BEAR 2002).

A report to the U.S. Congress by the Government Accountability Office (GAO) in December 2006 found that the policy option cited second most frequently by stakeholders as a top priority for federal action was to enact a national bottle bill. Respondents in the study said that the federal government would need to set its deposit amount sufficiently high to provide a measurable incentive for recycling, with 10 cents mentioned as the minimum amount necessary to accomplish this goal. The GAO also reported that retailers would likely oppose any redemption system that imposed significant additional costs on their operations, although support might be gained for an alternative redemption program, such as allowing consumers to return their used beverage containers to certified redemption centers (U.S. GAO 2006).

More Trash, Shrinking Landfills

During the Great Depression in the 1930s, most large communities established refuse collection programs and were trying various methods to dispose of the increasing amount of urban trash. One popular method was reduction, also known as the Merz process, which had been developed in the late 1880s and was used for organic wastes. Food, grass and yard clippings, and other materials were cooked for 8 to 10 hours and then compressed to make grease and fertilizer. But the process was expensive in comparison with dumping municipal trash into pits, so cities switched to landfills or incineration. Many of the waste sites were natural canyons, swamps, gullies, or abandoned mining pits that were virtually free for the taking. Garbage could be dumped into an earthen depression, compacted with earth-moving equipment, and then covered with soil. Sometimes chemical solutions were sprayed on the soil to kill insects or vermin.

This waste disposal option worked well into the mid-1980s, when the tipping cost paid by trash haulers to unload or "tip" their trucks into waste sites suddenly more than doubled. In Minneapolis, for instance, the tipping fee jumped from $5 to $30 per ton. Enforcement of the Resource Conservation and Recovery Act and one of its provisions, Subtitle D, led to the implementation of costly landfill safety standards under federal law. The problem was that landfills had routinely accepted, and commingled, hazardous waste with municipal solid waste, contaminating the soil and groundwater. By the late 1980s, so many landfills had been declared toxic that they comprised half of all Superfund sites (see below). Facilities were being forced to either clean up or shut down, with the majority doing the latter. The closures came at the same time that garbage output was exploding, putting pressure on the facilities that were still open (Rogers 2005).

Although some municipalities turned to incineration, that method of disposal, too, has problems. The emerging environmental movement had to make hard choices between the filthy air emissions of the combustors or the potential health hazards and environmental damage caused by landfills. Community opposition to both methods was widespread, as was the NIMBY (not in my backyard) syndrome. Recycling became more popular, but it was slow to gain acceptance as the amount of packaging waste increased faster than curbside recycling programs.

In 1988, nearly 8,000 landfills were operating in the United States, but that number began to decrease rapidly. By 1990, the number was about 6,300; by 1994, just over 3,500 were in operation; and by 1998, 2,300 facilities were operating. In 2005, the most recent year for which information is available, the total stood at 1,654, only about a fifth of the number of landfills that existed in 1988 (U.S. EPA 2006a). In response, local governments began looking for ways of expanding landfill capacity by considering other strategies, including enhanced recycling efforts.

One of the unanticipated consequences of the landfill closures is the effort to remake municipal landfills into parks and recreation areas, especially in parts of the United States where open land is at a premium. In most instances, the closed landfill sites are large mounds of dirt over yards of compacted trash. As the garbage below decays, some settling of the dirt above occurs, making the landscape less desirable for hard surfaces that could potentially crack. Most "reclaimed" landfills are also not used as football and soccer fields to avoid activities that could penetrate the cap on top. Cities are usually required to monitor the site for at least 30 years to make sure that the decaying garbage does not seep into the groundwater below. Any methane gas that is created from the buried trash is often burned off in an area that is not accessible to the public so it will not interfere with activities at the site.

Some of these reclaimed sites are small municipal dumps that are considered eyesores when their life as a landfill is over. In Chandler, Arizona, for instance, the planned Paseo Vista Recreation Area will sit on 2.2 million tons of compacted trash that is 38 feet deep. The original 64-acre city facility was opened in 1981 and closed in 2005, when its capacity was reached and the community's growth brought homes closer to the site. Even before it closed, voters approved $12.8 million in general obligation bonds to remake the landfill into a park, which will include a playground, picnic areas, equestrian trails, an archery range, and other amenities when it opens in 2009 (Fehr-Snyder 2007).

One of the most ambitious projects involves New York's Fresh Kills Landfill Complex, 2,200 acres of Staten Island that opened in 1948. Planners and designers have been examining the site for years in an attempt to determine how it could be reused after more than 50 years as a facility for dumping household waste. The western boundary lies along a waterway, and the landfill is part of a unique ecological area and watershed. The city conducted a two-stage international design competition to find innovative uses for the land and, in December 2001, selected three teams (two from the United States and one from the United Kingdom) whose proposals involved developing a park and recreational facilities while restoring the ecology of the site. The result is the Fresh Kills Lifescape, a 21st-century reclamation project that is expected to be completed by 2016. The four-part park will include a boat launch, meadows, an event lawn, marshes, and an activity center. This dramatic transformation will serve as a model of what can be done in other settings. As the landscape architect for the project notes, "It is the contemporary sense of healing the Earth as a technological notion" (Lange 2006).

Paying for the "Real" Cost of Waste

Most municipal governments and private companies that collect waste charge a fixed fee for picking up household trash, usually on a monthly basis. The cost may be part of a utility fee or charged separately. But environmental organizations argue that with a single, fixed fee, a household has no incentive to reduce or recycle its waste, as the volume to be collected has no bearing on the fee charged. The fee is the same whether the trash is separated into recycled waste and regular household waste or whether a household has one garbage can or five. In some communities, residents actually have to pay for the privilege of recycling trash because of extra fees charged for borrowing recycling containers or bins.

One way of providing an incentive for waste reduction is to implement "pay as you throw" programs, used to give residents options for managing household waste. Four basic types of these programs, also known as unit pricing, are in operation:

- Can systems allow customers to choose how many waste containers they want to set out for collection, with each can's size representing a size or weight limit, which then becomes the basis for collection fees.
- Bag systems require users to put their waste into a special bag with a distinctive color, tag, sticker, or logo that indicates it has been purchased at a store or recognized location. The customer prepays for the bag, which is guaranteed to be collected. The more bags the consumer

buys, the more they cost, serving as an incentive to reduce waste.

- Two-tier systems require payment of a flat fee for waste collection services, sometimes as part of a monthly charge or property tax, with one tier representing a basic cost for services and a second tier for waste collected beyond the basic service. Limitations may be imposed on how much trash can be collected each time, and fees may be based on volume.
- Weight-based systems assess charges contingent on the number of pounds of trash that are put out for collection. These programs are considered to be the most fair because they charge consumers for the amount of trash collected (with the weight based on the size of the container), rather than encouraging the compression of garbage into cans or bags (Tchobanoglous and Kreith 2002, 6.12).

Specialized Waste Management Challenges

Another major waste management issue relates to specialized disposal problems. For example, in the United States, communities routinely provide collection bins for items like newspapers, plastic, and glass. But electronic waste, or e-waste, is an emerging problem as computers, cell phones, and other electronic products become obsolete or stop working efficiently. California's Take-It-Back Partnership includes legislation that bans placing batteries, electronic devices, and fluorescent lightbulbs in regular trash. Instead, state and local governments are developing free, convenient ways for residents to recycle these items, as are cell phone companies. ReCellular, Inc., is the largest recycler and reseller of used wireless telephones and accessories in the industry, collecting, refurbishing, and reselling or recycling 3 million wireless handsets in 2004. It is a partner with the wireless industry's national organization, the Cellular Telecommunications and Internet Association, and has managed the group's national recycling foundation since its inception in 1999. The foundation works with more than 2,000 grassroots organizations and national charitable groups such as the March of Dimes and Goodwill Industries to collect wireless phones and support the groups' activities (ReCellular 2007). However, even if it is possible for e-waste like cell phones to be reduced or recycled on the domestic level, the problems electronic devices cause are not solved, as is illustrated in detail in Chapter 3 in the discussion on the international transport of e-waste.

Super Problems, Superfund

Chapter 1 provides definitions of the terms *toxic waste* and *hazardous waste,* two of the many types of refuse created in both residential and industrial settings. They include common household items such as cleansers, lawn care products, pesticides, and paint; waste created at manufacturing and mining operations; and refuse derived from the operations of other industries and companies such as metal plating facilities, chemical plants, furniture finishers, automotive dealers and gasoline service stations, dry cleaners, printers, and petroleum production facilities. Another type of waste is contamination that results from a slow leak from an underground storage tank or from an accidental spill.

Sites where toxic waste contamination is present pose numerous environmental and health concerns. Long-lived toxic substances, such as heavy metals like lead and mercury, and organic chemicals may be present for lengthy periods without degrading, creating problems even if no additional contamination is created. Chemicals may migrate through various pathways into groundwater or soil, or may bioaccumulate in the food chain. If they remain in the ground, they may not be exposed to the natural degradation that occurs when they are exposed to sunlight or oxygen. In addition, they may accumulate in environmental sinks, such as wetlands and river sediment.

The health impacts of many substances commonly found in hazardous wastes are not fully known or understood. Acute toxicity refers to effects that are intense and immediate, such as an individual being poisoned by ingesting a toxic substance. Some toxic waste can cause cancer even when the exposure is only a few parts per billion or less, as in the case of the waste generated and dumped by Hooker Chemical Company in New York. The owners operated a facility from 1942 to 1953 and legally buried 22,000 tons of chemical waste in canals on property they owned. Later, the company filled the canals with dirt and began to subdivide the land, which was purchased by developers, who built a residential neighborhood, Love Canal, New York. In the mid-1970s, residents began to complain of strange health problems, and researchers found an increased rate of cancer, birth defects, and spontaneous abortion, along with findings of cell aberrations. In August 1978, the government determined the area was unsafe and eventually evacuated more than a thousand households; it spent more than $30 million to purchase homes in the most contaminated areas. But as this example indicates, "the most difficult part of determining the toxicity of environmental contaminants is estimating the lifetime human health effects of exposure to low concentrations of toxins," which is the most common type of exposure from contaminated waste sites (Page 1997, 45).

The process of cleaning up hazardous waste is called remediation, which refers to a variety of actions that can be taken once a hazardous site is identified. Sometimes, cleanup involves containment in place, where a cap is placed on a site, such as a quarry where waste has been dumped, or a waste site may be fenced off and closed to the public. In some cases, the waste is excavated and removed to a secure facility off-site, usually where it is managed and then buried. This technique is often used when a large population is located near the site and the potential health risk is high. The term *treatment* denotes a way of attempting to remove or destroy the contamination and may involve thermal remediation (incineration to destroy organic compounds or to vaporize contaminants from mixtures of solids and water such as sewage sludge). Extraction, in contrast, uses filtration, ultraviolet, or ozone treatment or precipitation to remove contaminated waste. Recycling is possible when a concentration of a contaminant is sufficiently high in value to justify removing it from the site. Bioremediation and phytoremediation use microorganisms, such as bacteria, and plants, such as sunflowers, to degrade the waste and take the contaminated substances up through their roots, respectively (Page 1997).

Ideally, hazardous waste would be reduced or eliminated by the party responsible for creating it under the "polluter pays" principle. This is a legal tool whereby private parties that create hazardous waste must pay for any subsequent environmental cleanup of the site. The argument is that this principle deters additional contamination, and most people believe it is fair for polluting parties to bear the cost of cleanup, which is considered part of the cost of doing business. But incentives for companies to deal with waste in this manner appear to be insufficient, so the government instead operates under the "command and control" approach. This approach involves certain types of restrictions that regulate how a polluting company can operate, such as limiting the amount of hazardous waste that can be generated or requiring new factories and facilities to operate under the best technology available.

Another type of command-and-control restriction regulates toxic substances by controlling the entire life cycle of a product, such as a chemical, from "cradle to grave." In this approach, generators must determine whether substances are hazardous, based on qualities such as ignitability, corrosivity, reactivity, and toxicity. They must also obtain a firm identification number and permit for the generating facility. During shipment, this permit must accompany the materials in appropriate shipping containers and with a shipping manifest that tracks the material's movements from the time it leaves the generating facility until final disposal. Governments use this policy approach to minimize illegal disposal of hazardous wastes, sometimes referred to as "midnight dumping" (Page 1997, 61).

In the United States, the primary legislation dealing with hazardous waste cleanup is the 1980 Comprehensive Environmental Response, Compensation, and Liability Act (CERCLA), which established the Hazardous Substance Response Trust Fund, better known as Superfund. The initial objective of the law was to provide funding to clean up only the most seriously contaminated sites, arbitrarily set at 400 locations (based on the number of congressional districts), which were placed on the National Priorities List (NPL). The CERCLA legislation requires that the EPA analyze sites upon completion of the Hazard Ranking System screening and receipt of public comments. The list guides the agency in determining which sites warrant further investigation to assess the nature and extent of the human health and environmental risks associated with a site; identifying what remediation actions may be appropriate; notifying the public of affected sites; and serving as notice to potentially responsible parties that the EPA may initiate CERCLA-financed remedial action. The NPL is not, however, a statement of legal liability but rather serves an informational purpose.

The initial list of sites, announced in 1983, identified 406 locations; by June 2007, 1,243 sites were on the list, with another 61 sites proposed for listing, but only 319 sites had been deleted from the list (U.S. EPA 2007b). The trust fund was initially set at $1.6 billion, with revenues collected primarily from the assignment of the liability for cleanup costs. Specific liability stems from four classes of "potentially responsible parties":

1. The current owner or operator of a site
2. Any person who formerly owned or operated the site at the time of disposal of any hazardous waste
3. Any person who arranged for disposal or treatment of hazardous substances at the site
4. Any transporter of hazardous substances at the site

The question of who is liable and responsible for cleanup becomes even more complicated, however. A party can be strictly liable (extending the responsibility for the cost of cleanup beyond the requirement for culpable behavior), retroactively liable (extending to past actions, even if the activities were legal at the time they occurred), or jointly and severally liable (making any of the parties that can be identified liable for the entire cleanup costs or making the parties responsible for the proportional share of their contribution to the contamination event). Sometimes, the party with the greatest capacity to pay the cost of the cleanup, such as a large corporation with "deep pockets," will be held liable. When the responsible party is a local government, as is the case when the municipality operates a landfill, officials may be required to pay a much larger share because municipal solid waste (MSW) was the largest proportion of the total waste deposited at the site. In some cases, no one can determine how much of the waste that was contributed by each party was actually toxic (usually less than 1 percent of all MSW is thought to consist of hazardous material). In California, local governments have been sued for doing nothing more than issuing business licenses to private waste haulers who are paid directly by residents to pick up their trash (Page 1997, 83).

Although progress on cleaning up Superfund sites has been slow, there seems to be agreement that federal action is the only realistic way to make sure that cleanup occurs and public health is protected. Because other international priorities and issues have eclipsed most domestic concerns about the environment, the legislation establishing the Superfund program has lapsed and has not yet been reauthorized by Congress.

Radioactive Waste and Yucca Mountain

During World War II, the United States engaged in several research projects related to atomic power and nuclear reactors, resulting in the need to find a way to dispose of spent fuel. Some of the high-level radioactive waste was reprocessed, initially at the Hanford, Washington, site, where plutonium production reactors were operating as part of the Manhattan Project, and later, at West Valley, New York. In 1957, a committee from the National Academy of Sciences recommended that the government consider permanent disposal in deep salt formations, and in 1969, the government unsuccessfully experimented with an abandoned salt mine in Lyons, Kansas. Later proposals to develop a long-term retrievable surface storage facility were dropped due to opposition by environmental groups and the EPA. In 1970, the Atomic Energy Commission established a policy that the federal government would accept high-level waste for disposition (Cotton 2006).

Today, most of the reprocessed high-level waste that exists is part of national defense programs and is now stored in Idaho, South Carolina, and Washington State; the United States does not currently operate a reprocessing facility of its own. Other sources of spent fuel include nuclear power reactors that have been built across the country. Government officials have examined numerous storage alternatives for nuclear waste, including sending the waste into outer space, submerging it under ice caps and seabeds, and reprocessing the spent fuel, which is highly radioactive, into other types of fuel. The focus is now on two strategies: geological disposal (where the waste is buried for an indefinite period of time) and long-term storage at reactors where the waste is actually created. The decision about how and where to store nuclear waste has a long history that has pitted state and local officials and environmental groups against the federal government and the majority of Congress. As interest in nuclear power has been rekindled, the problem of what to do with the waste is receiving renewed attention.

The most controversial domestic project is the Yucca Mountain nuclear waste depository currently under construction in Nevada. The site is about 90 miles northwest of Las Vegas, at the edge of the Nevada Test Site, where nuclear weapons had been tested for years. A concerted effort to find a suitable place to store waste began in 1978, when President Jimmy Carter established the Interagency Review Group on Nuclear Waste Management. Four years later, Congress ordered the development of a permanent national disposal site for waste from commercial nuclear power reactors by enacting the Nuclear Waste Policy Act of 1982, directing the U.S. Department of Energy (DOE) to nominate at least five sites in a first round of repository selection and to recommend three of them to the president for further study. Initially, the DOE developed a list of nine sites; in 1986, the agency narrowed the list of potential sites to three: Yucca Mountain, Nevada; Deaf Smith County, Texas; and the Hanford facility in Washington State. A second round of recommendations examined 17 states in the eastern United States to mollify western state leaders who felt they had been singled out unfairly despite a requirement for a regional distribution of repositories. But the second round of selections was later cancelled, leaving the initial three locations for further study and evaluation.

After only a year, the Yucca Mountain site was designated as the only site that would be studied further under the provisions of the Nuclear Waste Policy Amendments Act. Plans were made to place up to 77,000 tons of high-level nuclear waste in permanent storage in a geologic repository. Currently, radioactive waste is being stored at 131 locations throughout the country. Officials estimated that the waste stored at Yucca Mountain would remain radioactive for a minimum of 10,000 years. Studies of the site's suitability, required under law, included mapping the mountain's geologic structure, collecting 75,000 feet of core samples, and collecting more than 18,000 geologic and water samples. More than six miles of tunnels were built to map the interior features of the locations where the radioactive material would be deposited. By 1989, the DOE concluded that because of their complexity, the studies were taking longer than anticipated, making it unlikely that the site would be available before 2010. The studies facility itself was not completed until 1996 (Cotton 2006).

Pressure to move more quickly developed in 1994, when utility companies sued the DOE because it was clear that the agency would not meet a 1998 deadline for accepting nuclear waste. A federal court judge sided with the companies and ruled that the government would be liable if it failed to meet the deadline. The Clinton administration vetoed a bill that would have allowed storage to begin after construction had been authorized, and most Democrats opposed the plan. But when George W. Bush was elected president in 2000, the mood in Washington, D.C., changed toward political support, except for representatives of Nevada, who began to mount strong opposition to any site in their state.

Proponents of the facility contended that a compelling national interest exists to complete the depository and that the facility is critical for national security and energy security as well. More than 161 million people live within 75 miles of one or more nuclear waste sites, which were intended to be temporary. The need to consolidate the waste in a single site deep underground was essential, they argued, in the face of the September 11, 2001, terrorist attacks. Because nuclear power provides 20 percent of the country's electricity, supporters noted that it was essential to find a site to secure the waste from one of the cheapest forms of power generation available. They criticized opponents for what they termed "scare tactics" even though radioactive materials have been transported for more than 30 years without any harmful release of radiation (Abraham 2002).

Opponents argued that Yucca Mountain is unsuitable because of questions related to the geologic integrity of the site, the potential that groundwater below will be contaminated, and the durability of the containers that would hold the waste. They sought to keep existing radioactive waste in storage where it is already located, near nuclear power plants. Many environmental organizations, in addition to their fears about environmental damage, felt that the risk of uncertainty was too great for such an experiment and that placing all the nation's radioactive waste in one state was inequitable.

The state of Nevada, which has no nuclear plants of its own, has gone to extraordinary measures to try to stop the facility from being operated in its backyard, using a variety of strategies. In early 2002, when the state was lobbying members of Congress to eliminate Yucca Mountain as a possible depository site, television advertisements were run in five states focusing on 10 senators whose votes on the project were considered to be uncertain: Missouri, Oregon, Utah, Vermont, and Wyoming. Environmental groups and a personal donation from the state's governor helped the state pay for the air time, combining the ads with an e-mail and postcard campaign and meetings with congressional staffers. Their primary argument was that nuclear waste shipments would travel through 43 states on highways and railroad tracks on their way to Nevada, with the shipments vulnerable to accidents or terrorist attacks (Grove 2002). Environmental groups called the planned waste shipments "mobile Chernobyl"—a reference to the 1986 nuclear disaster in Ukraine (CNN.com 2002). Nevada's congressional delegation also released a videotape showing a missile piercing a canister used to store radioactive material to show that transporting the waste is not safe.

Other anti–Yucca Mountain strategies were less politicized. In April 2002, for instance, the state engineer banned the Department of Energy from drawing water from state wells by invalidating a temporary permit that allowed the federal government to draw up to 140 million gallons of water a year. State officials hoped they might shut the project down by cutting off the water supply to the site. But the DOE simply switched to a newly built 1-million-gallon tank that would allow scientists to continue their work. The state had shut off the water once before in 2000, but the federal government successfully sued to keep the water flowing (Associated Press 2002).

In 2001, the DOE reported that no "showstoppers" arose in a scientific review of the Yucca Mountain site, which at the time was estimated to cost $58 billion over 10 years for construction, operation, and monitoring. Secretary of Energy Spencer Abraham approved the site, noting that the project had benefited from 24 years of scientific study, at a cost of $4 billion, showing that Yucca Mountain was suitable and safe. The 1982 Nuclear Waste Policy Act allowed the governor of Nevada to veto the president's decision, which he did in April 2002. Gov. Kenny Guinn cited three reasons for the state's opposition.

> Nevada strongly opposes the designation of Yucca Mountain for nuclear waste disposal because the project is scientifically flawed, fails to conform to numerous laws, and the policy behind it is ever-changing and nonsensical. The Department of Energy has so compromised this project through years of mismanagement that Congress should have no confidence in any representation made by DOE about either its purpose or its safety. (Guinn 2002)

Nevada officials called the Yucca Mountain project "legally dead" because of the governor's veto. The mayor of Las Vegas was less sanguine about the state's chances. "If we don't prevail in the House, we will prevail in the Senate. And if we don't prevail in the Senate, we will definitely prevail in the courts because might is right" (Neff 2002). But only a few weeks later, the U.S.

House of Representatives voted 306 to 117 to override Nevada's attempts to stop it from moving forward through a veto override. The Senate, which was under a July 26 deadline to act, also voted to override the state's objections and protests in its 60 to 39 decision to proceed to a voice vote.

President Bush subsequently signed legislation designating Yucca Mountain as the government's preferred site, noting the administration would seek an operating permit from the Nuclear Regulatory Commission (NRC). The state filed several lawsuits in federal appeals court challenging various rules issued by federal agencies related to the facility. Nevada also filed a petition with the NRC seeking rule changes to its licensing procedures that would require the DOE to prove the site will be safe and that geology, rather than waste containers, would be the main protection against leaking. Thus far, the NRC has not made a final decision on whether the DOE will be allowed to proceed with construction and then licensed to operate the Yucca Mountain facility. The NRC must conclude that the repository will meet its reasonable expectation that the safety and health of workers and the public will be protected. No date for an announcement of the decision has been set.

Debris from Disasters

The volume of waste resulting from natural and human-caused disasters can be virtually impossible to comprehend. Two major disasters in the United States created debris that called for immediate removal responses that were unprecedented. Sometimes, the desire to deal with the problem immediately has led to solutions that may have unintended consequences later on.

The human tragedy of the September 11, 2001, terrorist attacks on the World Trade Center in New York City is a haunting memory that cannot be erased from American society's collective memory. But in order to move beyond the physical damage at the site, officials had no choice but to deal with the waste management nightmare of 1.2 million tons of building debris from the center complex. The Federal Emergency Management Agency (FEMA) and the U.S. Army Corps of Engineers worked jointly with the City of New York in dealing with debris removal, a task made more difficult because of the need to search for human remains. The debris was also contaminated with jet fuel from the two airliners that crashed into the buildings, as well as petroleum, Freon, and other chemicals. The 12-square-block site was different than most disaster areas, where the two agencies have typically dealt with downed trees and residential debris caused by natural disasters like floods, earthquakes, and hurricanes. This type of waste management posed numerous challenges. Initially, officials had to decide where to take the debris and how to get it there. This decision making was complicated by the fact that the entire site was considered a crime scene because it was attacked by terrorists.

A decision was made to take the debris to the Staten Island landfill, a 2,200-acre facility also known as the Fresh Kills Landfill Complex (discussed above). The city's Department of Sanitation and the New York Police Department took on responsibility for managing the Fresh Kills facility, which had been closed in March 2001, about 20 miles by roadway from the World Trade Center. FEMA funded the removal project, while the Army Corps of Engineers developed the debris management plan; helped develop contract specifications; monitored the removal project operations; and facilitated the loading of 240 trucks, 70 barges, and more than 250 pieces of heavy equipment (Stroupe 2001).

Because the debris pile consisted of urban building materials, decisions had to be made about what could be recycled or salvaged from the site. The estimated 300,000 tons of structural steel that was recovered was sold by the city and sent to two recycling yards, where it was later sold to scrap companies for about $120 per ton, considered a reasonable fee in comparison with the usual $150 per ton rate for scrap steel. About 50,000 tons of steel was purchased by a Shanghai company to be melted down and reprocessed into new items such as kitchenware and other household goods (China.org.cn 2002). The rest of the debris went to Fresh Kills for sorting and examination, including extensive searches for human remains. Controversies over the thoroughness of the searches continue.

When hurricanes Katrina and Rita ravaged the U.S. Gulf Coast in 2005, they left behind an inestimable loss of human life and environmental damage. Of immediate concern to officials seeking to expedite the region's recovery was the demolition of damaged houses and other buildings, along with the removal of other types of waste generated from massive flooding after the storm surge. As was the case with the World Trade Center disaster, FEMA asked the Army Corps of Engineers to perform debris removal in September 2005 on behalf of various cities and parishes, including New Orleans. The Corps then contracted with various vendors to conduct the cleanup and to take the waste to properly permitted landfill sites. FEMA provided funds for eligible debris removal costs through December 21, 2006; after that time, the city would incur 10 percent of the costs, providing an incentive for a timely cleanup process.

In Louisiana, the Department of Environmental Quality (DEQ) estimated that more than 95,000 houses would need to be torn down as a result of hurricane damage and flooding, with the demolition resulting in more than 20 million cubic yards of waste in Orleans Parish. It is almost impossible to imagine how local officials might deal with handling this amount of debris, especially because the government infrastructure had been destroyed or damaged and officials were unable to mobilize employees and equipment to speed recovery. There was also a desire to expedite the process, as recovery would be dependent, in large part, on how quickly the waste could be removed so rebuilding could begin. More than 400 sites were approved for the staging, chipping, grinding, burning, and disposing of hurricane debris within Louisiana (Louisiana DEQ 2006).

Although New Orleans had regional landfills that could accommodate some of the debris, most were designed to handle hazardous materials, not demolished houses. Normally, the process of siting and opening a landfill at the edge of a major city would be a lengthy process punctuated with years of planning, public hearings, and environmental reviews to determine whether potential hazards or negative impacts might arise. Mayor C. Ray Nagin, who had invoked emergency powers after Katrina to waive the city's zoning ordinances, determined that New Orleans did not have the time and resources to develop a new site, so instead, the city converted a deep pit amid the wetlands east of the city into a landfill in April 2006. The pit had been created when construction companies excavated sand to use in their projects. Officials received emergency authorization under the federal Clean Water Act to have the pit accept the waste, with a potential for taking in as much as 6.5 million cubic yards of debris.

The Chef Menteur landfill sits across from a canal and the largest urban wildlife refuge in the United States, the 23,000-acre Bayou Savage, which is used by migratory birds. It is also less than a mile from a community of 1,000 Vietnamese American families in the Village de l'Est neighborhood who were rebuilding after the storm. Not unexpectedly, opposition to the landfill operation stemmed from local residents who worried about toxins that might leach out from the pit and from environmental organizations that voiced similar fears about the creation of the dump. An attorney who worked with the Louisiana Environmental Action Network to oppose the landfill noted that it made "a second disaster out of the first one" (Curtis 2006). A local Catholic priest said the mayor's decision to place a landfill in the neighborhood would hurt the community's rebuilding efforts. "This will have a chilling effect on our recovery. It's certainly a black eye to us as the people trying to recover. We thought we could reason with them to show them all the compelling reasons not to put a landfill in a wetland area. There seems to be a disregard for human safety as well as recovery" (Russell 2006).

Unlike sanitary landfills, whose sides and bottom are lined with clay for another barrier layer to prevent the leaching of toxins into the water table, the Chef Menteur site was unprotected and did not have a leach detection system installed when it opened. Officials for the operator noted that the site was naturally clay lined and had been used successfully for years as a landfill for household waste. Another unlined landfill that had been closed in 1986, Old Gentilly, had also been hastily reopened for demolition waste. One of the serious problems related to the demolition and construction waste was the sorting of hazardous materials from hurricane-damaged sites. The order creating the landfill allowed everything within the four walls of a house to be dumped into the pit, and that debris included cleansers, pesticides and paints, petroleum products, degreasers and solvents, and any other materials inside. Although spotters at the landfill were supposed to look for hazardous materials as trucks full of debris unloaded, only some of those items were likely to be diverted, officials warned. As rain fell on the site, leaching chemicals mixed with the water and would eventually flow into the canals in the city.

Leaders within the Vietnamese community argued that the site had been selected because New Orleans officials anticipated little opposition to the dumping, but local residents framed the issue as a question of environmental justice, gaining the media spotlight. Both the state and the Army Corps of Engineers had argued that if the dump was not used, the cleanup of the city would take much longer, and waste would have to be shipped farther away. One official noted that the community's opposition did not take into consideration the scale of the emergency. "Some people can't seem to understand that this is not business as usual" (Eaton 2006).

The mayor had an additional incentive to move quickly to open the dump site. The company hired to operate it, Waste Management, Inc., pledged to give 22 percent of the revenue it received from tipping fees back to the city, estimated at $860,000. A review of the agreement conducted by the U.S. Department of Homeland Security, completed late in the decision-making process at the request of a member of Congress, found that the facility was not the most cost-effective landfill in the region and that Waste Management's tipping fees were significantly higher than other operators. The report found that the landfill operator had passed the cost of the "donations" to the Corps' contractors, which were then passed along to FEMA. The agency did not require New Orleans to pay for debris removal from the two hurricanes, but the report noted that the city should not profit from the operations (U.S. DHS 2006).

Originally, the Chef Menteur landfill was expected to operate for 12 months, but it was closed in August 2006, ostensibly because the landfill operator had failed to apply for a conditional use permit. Waste Management subsequently filed a lawsuit in federal court to reopen the facility, but in an October 2006 decision, a civil district judge ruled against the company, saying that it should have applied for the permit to operate the landfill on a long-term basis well before the closure order had been issued by Mayor Nagin.

Although these two cases are at the extreme end of the waste management problems faced by government officials, they demonstrate an increasing awareness of the battle over landfill use from various perspectives. In addition to the NIMBY battles of residents who do not want facilities located nearby, environmental concerns factor into every siting decision. Even after emergencies like the World Trade Center attacks and the Gulf Coast hurricanes, decisions on how to deal with debris become complex and highly politicized.

Siting Waste Management Facilities

Waste management typifies the NIMBY, NIMTOO (not in my term of office), and BANANA (build absolutely nothing anywhere near anything) syndromes arising during times of political decision making about siting waste facilities. Even when government officials have efficient technologies to work with, such as WTE plants, they often are reluctant to initiate projects because of a fear of voter backlash. The founder of a Texas-based company that owns landfills throughout the eastern part of the United States put it this way: "Landfills in the United States are not environmental issues. They are strictly political" (Associated Press 2005).

Operational issues associated with waste management have created a multitude of problems but few solutions, especially where landfills are concerned. One of the issues is the odor and air pollution that are produced, either from the trash and organic matter itself or during collection. In urban areas, for instance, residents often complain about the smell that emanates from waste transfer stations, described in Chapter 1. In New York City, transfer stations are a way for collection trucks to "tip," or unload, residential and commercial trash until it is picked up by larger trucks. A 2004 study by the nongovernmental organization Environmental Defense found that transfer facilities were being placed in a small number of New York City's neighborhoods, where hundreds of trucks moved through the area each day. The odors from the garbage are coupled with the diesel exhaust from the trucks, especially as they idle while waiting to unload. Residents nearby have complained not only about the smells but also about the increases in respiratory illnesses such as asthma (Cruz, Outerbridge, and Tripp 2004).

New York City mayor Michael Bloomberg has proposed closing the residential waste transfer stations and reopening four city-owned marine transfer stations so that the garbage would be loaded onto barges and sent to an incinerator facility in New Jersey or to landfills in Virginia and Pennsylvania. The facilities already exist but were closed in 2001, when New York stopped using the Fresh Kills landfill on Staten Island except for the World Trade Center debris. The stations will need to be revamped at a cost estimated at $85 million each. The problem is not just the cost involved, however. A New York City study found that 91 percent of the trash headed for landfills is controlled by one of three companies, raising questions about trash monopolies. New York's residents have also refused to support building a landfill in their own state and voted down a proposal, eventually abandoned in the 1990s, to build an incinerator in the Brooklyn Navy Yards, both now considered "unspeakably bad ideas" (Brustein 2004).

Some progress is being made to deal with the city's trash problems. In April 2007, the eight-mile-long Staten Island Railroad was reopened for the first time since 1991. It allows garbage to be transported directly to sealed containers on railcars. City officials said the reopening of the line would cut out thousands of truck trips on Staten Island each year, easing traffic, reducing air pollution and odors, and minimizing the impact on communities (Rivera 2007).

But when landfills close in big cities like New York, the trash has to be sent somewhere, and that "somewhere" is often far away from the businesses and residences where it was created. Long hauls from urban centers to rural areas are now commonplace, and the Congressional Research Service estimated in 2003 that one-quarter of all municipal trash in the United States crossed state lines for disposal. Ten states imported at least 1 million tons of trash in 2003, up from only two states in 2001. New York, for instance, transports more than 1,300 tons of garbage each day to Fox Township, Pennsylvania, about 130 miles north of Pittsburgh. Virginia imported an estimated 7.8 million tons of garbage in 2005. Some communities get financial benefits from charging waste haulers tipping fees for every ton of garbage that is brought in; in Fox Township, for instance, the fees for importing garbage have enabled officials to purchase new police vehicles and fire trucks in a community of less than 4,000 people (Associated Press 2005).

Waste haulers and handlers have focused on rural areas for several reasons. Many rural communities are experiencing economic decline, reductions in business activity, and population decreases. New projects, such as incinerators, represent the promise of increased levels of employment, but they also represent the problems associated with providing government services and the potential for immigration, or the act of waste projects being sited in areas far from where the waste is generated. The project facilities that generate most of the new public-sector revenues may be located in one county, for instance, while most of the project-related people live in a different school district or county. "In general, the effects are most positive for those whose interests [through business sales or employment] are most directly tied to the development. They are least positive for those who have no direct link to the development" (Murdock, Krannich, and Leistritz 1999, 37).

One study of the impacts of waste projects in 15 communities in five states in the western United States, however, found little indication that waste-affected communities have experienced extensive economic growth as a result of waste projects or that the fiscal conditions of the areas have been substantially enhanced due to such projects. Nor did the study find any indication that the use of government services increases or that the population declines in waste-affected rural communities. But a substantial difference was seen in rural areas between those who support and those who oppose siting a waste facility, the study reports. Leaders, who are older, are better educated, have higher incomes, and are more likely to own property, are more likely to support a facility being built, in comparison with residents (Murdock, Krannich, and Leistritz 1999).

The challenges of siting landfills have led to controversial exchanges and debates, with some waste management companies taking extreme measures to gain approval for projects by offering incentives to decision makers. For example, in 1990, the Halle Companies filed requests to build a rubble landfill on property the firm owned in Maryland's Chesapeake Terrace area. The requests were denied by local officials and appealed by the company over the next three years, when a county board of appeals finally granted approval with a series of conditions related to road improvements, hours of operation, fencing, and a provision that the landfill could only be operated for 12 years. Local residents and organizations appealed the proposed project, and in 1994, the decision was reversed by the circuit court. That decision was then overturned by the court of appeals, and over the next several years, the matter wound its way through more administrative and judicial hearings. In 2002, the company, now called National Waste Management, made an agreement with the Delaware Nation of Oklahoma to take over the property where the landfill was planned, placing its operations under the federal government and the Bureau of Indian Affairs. Because the federal regulations on landfills on Native American lands are much less strict than those of the state of Maryland, the permitting process was assumed to be much easier.

Two years later, two civic organizations that had originally been a part of the opposition to the facility agreed to support the Chesapeake Terrace landfill after the company agreed to donate money to community organizations, set up ball fields, and build a new high school for the county. Another local organization had successfully negotiated the building of a swimming pool and clubhouse and a donation to a high school athletic booster club, while a third group, made up of residents closest to the proposed landfill, continued its opposition. In 2007, one of the groups that had pledged its support withdrew from the agreement, with one leader calling it "not worth the paper it is printed on" (Stewart 2007). The company continues to pursue the project, estimated to cost $560 million, with an expectation that construction would take 15 to 20 years to complete.

Is Help on the Way?

New technology, policies, and changing attitudes may be on the horizon in many communities and in industries grappling with waste management challenges. Some solutions rely on partnerships between government, nongovernmental organizations, and private-sector businesses, while others represent advances in how waste is reduced at the source or how it is handled, or changes in how people feel about trash and recycling.

The problems of dealing with hazardous waste, described earlier, have led to several sector-based initiatives with the support of the EPA. Sites that once might have been considered inappropriate for further development or use are now being reclaimed. Brownfields, for instance, are areas within a community where potential land use is complicated by the fact that a hazardous substance, pollutant, or contaminant is present or potentially present. The EPA works with state and local governments and tribal nations to provide grants to clean up and revitalize brownfields to turn them into productive community use. Since the beginning of the brownfields program, the EPA has awarded more than a thousand assessment grants totaling more than $262 million, 217 revolving loan fund grants totaling $202 million, and 336 cleanup grants totaling $61 million. Other grant programs provide training and technical assistance. In addition, federal tax incentives are available to allow environmental cleanup costs to be deducted from a company's yearly tax liability (U.S. EPA 2007a).

A typical example of a Superfund brownfields project is the Solitron Microwave Site in Port Salerno, Florida. The 20-acre site had been used for plating and manufacturing processes from

1963 to 1987. After the facility was shut down, local health department officials discovered 12 privately owned local wells in which contaminant levels were higher than federal drinking water standards allow, and in 1998, the EPA added Solitron to the NPL. The Army Corps of Engineers, working with the EPA, excavated soil near the rear of the building, and as a part of redevelopment, the developer demolished the building. Additional hazardous soil was excavated from beneath the building. The city now has a 20-acre industrial park instead of an abandoned facility considered an eyesore to the community, and an additional 150,000 square feet of warehouse and light industrial space to revitalize the community by providing new jobs and cleaning up an old, dilapidated building (U.S. EPA 2007c).

Similar sector-based initiatives are underway to clean up and reuse other types of sites that are similar to brownfields that may have had different types of uses. Portfields, for instance, are a type of brownfield located in port communities. Three cities were chosen for a pilot program for cleanup in 2003: Bellingham, Washington; New Bedford, Massachusetts; and Tampa, Florida. The program helps communities assess environmental and health concerns, leverage resources to revitalize waterfront areas, improve marine transportation, and protect and restore critical habitat. Similarly, the federal programs designed to reclaim railfields, mine-scarred lands, and areas where waste soil from underground storage tanks exist help restore areas through an interagency approach.

Agencies are also turning to innovative technological strategies to deal with various types of waste. Among the newer efforts is the use of phytoremediation—the direct use of green plants and their associated microorganisms to stabilize or reduce contamination in soils, sludges, sediments, or water. The process has been used at more than 200 waste sites nationwide since the early 1990s. The plant species that are used are selected on the basis of their ability to extract or degrade the contaminants in the waste, sometimes through their root systems. An example of the use of phytoremediation is the Oregon Poplar Site in Clackamas, Oregon. This site is a three-acre, abandoned grassy field in an area that was primarily designated for commercial and light industrial use. Contaminants that included volatile organic compounds had been illegally dumped there, and the area was particularly important because it paralleled a small stream. Although no risk to human health was evident because of the low level of contamination, the groundwater beneath the site is shallow and in hydraulic contact with the stream. Hybrid poplar trees were planted there as part of an effort to phytoremediate the area in 1998, and four years later, the trees not only had survived but had shown considerable growth. Tissue samples taken from the roots of four of the larger trees indicated that they were actively removing the chemical contamination from the waste soil (U.S. EPA 2007d).

The EPA has developed several other innovative programs, many of which involve public outreach and education. WasteWise, for instance, encourages companies to conduct a waste assessment to help employees identify waste-reduction strategies and establish a baseline for measuring progress. Participants, called partners, include large and small businesses; hospitals; universities; nonprofit organizations; and state, local, and tribal governments. The EPA provides free technical assistance and coordinates regional and national events to help partners network to find solutions. The program is free, voluntary, and, the EPA says, flexible and cost effective.

Some of the smaller-scale programs are inexpensive and easily replicated, such as the Schools Chemical Cleanout Campaign. The goal of the program is to ensure that schools are free from the hazards associated with mismanaged chemicals, which are found in sources such as maintenance facilities, chemistry labs, and nurses' offices on school grounds. While concern exists about the risk from spills, fires, and other accidental exposure caused by chemical accidents, an awareness also exists that these problems lead to school disruptions and closures, as well as thousands of dollars spent in repairs. The EPA program provides administrators and parents with the resources needed to develop a successful chemical management program that also involves chemical suppliers and waste handlers.

Technological Advances

As this chapter illustrates, the problem of waste management includes not only reducing the amount of waste created and developing strategies to increase the amount of waste that is recycled but also finding ways to deal with waste that must be treated. Technological advances are having an enormous impact on waste management treatment options, although the solutions do not fit every type of waste. For instance, St. Lucie County, Florida, announced in 2006 that it was planning a 100,000-square-foot facility that would allow it to close its dump, generate electricity, and help build roads by using lightning-like plasma arcs to vaporize garbage. The facility would be privately built and would be the largest of its kind in the world if plans to operate it by 2011 are finalized. It would vaporize 3,000 tons of trash per day and would incinerate trash from the existing dump as well, allowing officials to close it within 18 years. Synthetic, combustible gas produced by the plant would run turbines to create electricity that would be sold, along with steam that would power a neighboring juice plant. The sludge from the county's wastewater treatment facility would also be vaporized and then sold for use in road and other construction projects. The company would absorb the entire cost of building and operating the $425 million facility, at no cost to the county (Associated Press 2006).

Similar changes are affecting other waste operations. Throughout most of U.S. history, the debris from logging and milling operations has been burned as useless waste. But using biomass as fuel, especially woody biomass such as small tree limbs and sawdust, is now the number one form of renewable energy in the United States. Although biomass contributes only about 3 percent of the nation's energy needs, wood waste can be burned to generate steam and electricity. Biofuel development is becoming especially important in the U.S. West as a way of utilizing the waste created as a result of thinning forest vegetation and trees to reduce the danger of wildfires, especially in the wildland-urban interface. The U.S. Forest Service has made thinning and forest restoration a major priority, but until recently, little economic value has been realized in the woody biomass that results from new policies such as the Healthy Forests Restoration Act of 2003. Aggressive goals for hazardous fuels reduction are only now being coupled with finding ways to reduce the country's reliance on fossil fuels (U.S. Department of Agriculture 2008).

Most of the woody biomass applications involve small geographical areas, usually in rural communities where wood chips are readily available. The Forest Service has created a program called Fuels for Schools to encourage schools to replace heating systems that burn fossil fuels with those that use wood chips. The program has more than a dozen projects operating in four states, with others in the design or construction phase. In addition to its use in rural schools, woody biomass heating is being used in a correctional facility, in a hospital, and at the University of Montana's campus in Dillon. In states like Minnesota, woody biomass is used as an industrial-scale fuel to replace coal; in research programs, the cellulose in forest debris is being studied to determine whether it can be turned into liquid fuel similar to ethanol. These projects utilize material that would otherwise be sent to a landfill or burned on the ground (Bryant 2007).

Consumers can also make changes, many of which involve lifestyle changes but some of which are oriented toward innovative practices. In 2003, the practice of "freecycling" was started as a way of helping individuals find new uses for unwanted items. The first program began with Deron Beal of Tucson, Arizona, who was working for a nonprofit organization, Downtown Don't Waste It, a recycling program that collected unwanted office furniture and supplies. The group was hoping to reduce the amount of materials going to local landfills or being dumped in the desert. Other nonprofit organizations he was working with were willing to take some items, such as computers, and others needed items such as office supplies. He decided to start a computer listserv to bring the parties together, and then he expanded it to cover any free item anyone wanted to just give away. He sent out a notice on May 1, 2003, starting out with a posting for a queen-size bed he no longer needed. As the project grew, individuals started listing items they needed as well as ones they wanted to give away, and an article in the local newspaper helped add more members to the group. Various environmental organizations began publicizing freecycle groups, which began to spread across the country. The listservs have a moderator and a handful of rules, such as keeping all items free, legal, and appropriate for all ages. They discourage trading, politicking, and off-topic e-mail messaging. Most groups suggest that nonprofit organizations be the first ones to be offered items, but supporters note that freecycling often finds homes for items that charitable groups do not want. Now, nearly 4,000 local freecycling groups are in operation, with a membership of more than 3.5 million people in countries around the world (Nijhuis 2004).

The system is not a perfect solution. The group uses as its tagline "Changing the world one gift at a time," but some critics believe the organization sold out to greed when it accepted two grants of corporate support of $230,000 from the trash collection company Waste Management, Inc., in 2005. Additional criticism came from a deal to accept advertisements for the search engine Google on the network's Web site. The original group has also been criticized for staunchly defending its trademark and for imposing unnecessarily strict rules on members (Walker 2007).

What is abundantly clear is that waste management problems will not be solved by any single solution. Public officials must work cooperatively with waste haulers, who must work with the public. Government regulators must deal with industries and business organizations in order to develop rules that will meet the needs of both. Citizens, as consumers, will need to rethink their behaviors and attitudes about the trash they produce and must take responsibility for what becomes of each item. As long as waste management stays near the bottom of the political agenda, however, there are few incentives for new solutions or change.

References

Abraham, Spencer. 2002. "One Safe Site Is Best." *Washington Post,* March 26. [Online article; retrieved 4/18/02.] http://pqasb.pqarchiver.com/washingtonpost/access/111518545.html?dids=111518545:111518545&FMT=ABS&FMTS=ABS:FT&fmac=&date=Mar+26%2C+2002&author=Spencer+Abraham&desc=One+Safe+Site+Is+Best.

Associated Press. 2002. "State Shuts Off Water to Yucca Mountain; Research Continues." KVBC.com, April 10. [Online article; retrieved 4/18/02.] http://www.kvbc.com/Global/story.asp?S=738840.

Associated Press. 2005. "As Landfills Close in Big Cities, Garbage Travels Farther." *USA Today,* July 12. [Online article; retrieved 6/1/07.] http://www.usatoday.com/news/nation/2005-07-12-trash-diaspora_x.htm?loc=interstitialskip.

Associated Press. 2006. "Florida County Plans to Vaporize Landfill Trash." *USA Today,* September 9. [Online article; retrieved 4/17/07.] http://www.usatoday.com/news/nation/2006-09-09-fla-county-trash_x.htm.

Brustein, Joshua. 2004. "The Mayor's New Garbage Plan." *Gotham Gazette,* October 18. [Online article; retrieved 4/11/07.] http://www.gothamgazette.com/article/20041018/200/1151.

Bryant, Chris. 2007. "Fuels from the Forests." *Forest Magazine* (Spring): 33–37.

Burros, Marian. 2007. "Just the Thing to Carry Your Conscience In." *New York Times,* July 18. [Online article; retrieved 10/7/07.] http://www.

nytimes.com/2007/07/18/dining/18bags.html?_r=1&scp=1&sq=%22 just+the+thing+to+carry+your+conscience+in%22&st=nyt&oref=slogin.

Businesses and Environmentalists Allied for Recycling (BEAR). 2002. *Understanding Beverage Container Recovery: A Value Chain Assessment.* [Online report; retrieved 7/7/07.] http://www.globalgreen.org/bear/Projects/index.html#files.

California Integrated Waste Management Board. 2006a. "Pressure-Sensitive Adhesives: A Sticky Recycling Problem." [Online article; retrieved 7/9/07.] http://www.ciwmb.ca.gov/BizWaste/OfficePaper/PSAFacts.htm.

California Integrated Waste Management Board. 2006b. "What Is a Zero Waste California?" [Online article; retrieved 2/7/07.] http://www.zerowaste.ca.gov/WhatIs.htm.

Carpet America Recovery Effort (CARE). 2007a. *Annual Report.* [Online report; retrieved 7/10/07.] http://www.carpetrecovery.org/annual.php.

Carpet America Recovery Effort (CARE). 2007b. "About CARE." [Online information; retrieved 4/16/07.] http://www.carpetrecovery.org/about.php.

Center for Policy Alternatives. 2007. "Bottle Bill." [Online article; retrieved 7/7/07.] http://www.stateaction.org/issues/issue.cfm?issue=BottleBill.xml.

China.org.cn. 2002. "Baosteel Will Recycle World Trade Center Debris." China Internet Information Center. [Online article; retrieved 6/24/07.] http://www.china.org.cn/english/2002/Jan/25776.htm.

CNN.com. 2002. "Senate Approves Yucca Mountain Nuclear Waste Site." [Online article; retrieved 7/21/02.] www.cnn.com.

Conservatree. 2007. "Source Reduction: Using Your Buying Power." [Online information; retrieved 7/11/07.] http://www.conservatree.org/paper/Choose/SRpurchasing.shtml.

Container Recycling Institute. 2007. "Bottle Bill Resource Guide." [Online information; retrieved 7/7/07.] www.bottlebill.org.

Cotton, Thomas. 2006. "Nuclear Waste Story: Setting the Stage." In *Uncertainty Underground: Yucca Mountain and the Nation's High-Level Nuclear Waste,* edited by Allison M. Macfarlane and Rodney C. Ewing, 29–44. Cambridge, MA: MIT Press.

Cruz, Ramon J., Thomas Outerbridge, and James T. B. Tripp. 2004. *Trash and the City: Toward a Cleaner, More Equitable Waste Transfer System in Manhattan.* New York: Environmental Defense.

Curtis, Wayne. 2006. "A Heap of Sorrows." *Grist,* August 10. [Online article; retrieved 6/1/07.] http://www.grist.org/news/maindish/2006/08/10/curtis/index.html.

De Young, Raymond, Sally Boerschig, Sarah Carney, Anne Dillenbeck, Mark Elster, Susan Horst, Brad Kleiner, Bruce Thomson. 1995. "Recycling in Multi-Family Dwellings: Increasing Participation and Decreasing Contamination." *Population & Environment*, January. [Online article; retrieved 4/29/07]. http://www.springerlink.com/content/81151726u10769ul/?p=3dc0eeb6f52b41caa6976d488cd2c0cf&pi=0.

Eaton, Leslie. 2006. "A New Landfill in New Orleans Sets Off a Battle." *New York Times*, May 8. [Online article; retrieved 6/24/07.] http://www.nytimes.com/2006/05/08/us/08landfill.html?scp=1&sq=%22a+new+landfill+in+new+orleans%22&st=nyt.

Fehr-Snyder, Kerry. 2007. "Chandler Aims to Turn Closed Dump into Park by 2009." *Arizona Republic*, September 24, B1.

Film and Bag Federation. 2007. "Environmental Issues." [Online information; retrieved 4/26/07.] http://www.plasticsindustry.org/about/fbf/environment.htm.

Gitlitz, Jennifer, and Pat Franklin. 2007. *Water, Water Everywhere: The Growth of Non-Carbonated Beverages in the United States.* Washington, DC: Container Recycling Institute.

GrassRoots Recycling Network. 2002. "Reduce, Reuse, Refill!" [Online article; retrieved 7/7/07.] http://www.grrn.org/beverage/refillables/index.html.

Grove, Benjamin. 2002. "Yucca Ads Focus on a Handful of States." *Las Vegas Sun*, April 17. [Online article; retrieved 4/18/02.] http://www.lasvegassun.com/news/2002/apr/17/yucca-ads-focus-on-a-handful-of-states/.

Guinn, Kenny. 2002. "Statement of Reasons Supporting the Governor of Nevada's Notice of Disapproval of the Proposed Yucca Mountain Project." April 18.

Gurnon, Emily. 2003. "The Problem with Plastics." *North Coast Journal Weekly*, June 5. [Online article; retrieved 7/9/07.] http://www.northcoastjournal.com/060503/cover0605.html.

Halwell, Brian. 2007. "Good Stuff? Plastic Bags." [Online article; retrieved 4/26/07.} http://www.worldwatch.org/node/1499.

Lange, Alexandra. 2006. "Fresh Kills in 2016." *New York Magazine*, June 5. [Online article; retrieved 4/11/07.] http://nymag.com/realestate/features/2016/17149/.

Lapidos, Juliet. 2007. "Will My Plastic Bag Still Be Here in 2507?" *Slate*, June 27. [Online article; retrieved 7/11/07.] http://www.slate.com/id/2169287/.

Louisiana Department of Environmental Quality (DEQ). 2006. "Letter to Mayor Ray Nagin Regarding the Chef Menteur C & D Disposal Facility." July 21. [Online article; retrieved 7/1/07.] www.deq.louisiana.gov.

Minnesota Office of Environmental Assistance. 2007. "A National Agreement on Carpet Recycling." [Online information; retrieved 7/10/07.] www.pca.state.mn.us/oea/carpet.

Moore, Charles. 2003. "Trashed." *Natural History* (November). [Online article; retrieved 4/11/07.] http://nhmag.com/search.html?keys=%22trashed&sitenbr=157877211&bgcolor=%23C7E0B0.

Murdock, Steven H., Richard S. Krannich, and F. Larry Leistritz. 1999. *Hazardous Wastes in Rural America: Impacts, Implications, and Options for Rural Communities.* Lanham, MD: Rowman & Littlefield.

Neff, Erin. 2002. "Yucca Project Believed to Be 'Legally Dead.'" *Las Vegas Sun,* April 9. [Online article; retrieved 4/18/02.] http://www.lasvegassun.com/news/2002/apr/09/yucca-project-believed-to-be-legally-dead/.

Nijhuis, Michelle. 2004. "Give It Away, Give It Away, Give It Away Now." *Grist,* May 17. [Online article; retrieved 7/11/07.] http://www.grist.org/news/maindish/2004/05/17/nijhuis-freecycle/index.html.

Page, G. William. 1997. *Contaminated Sites and Environmental Cleanup.* San Diego: Academic Press.

ReCellular, Inc. 2007. "About Us." [Online information; retrieved 6/27/07.] http://www.recellular.com/about/index.asp.

Reusable Bags.com. "The Numbers . . . Believe It or Not." [Online article; retrieved 7/11/2007.] http: www.reusablebags.com/facts.

Rivera, Ray. 2007. "City Reopens Old Rail Line for Staten Island Trash and Freight." *New York Times,* April 18. [Online article; retrieved 4/19/07.] http://www.nytimes.com/2007/04/18/nyregion/18staten.html?scp=2&sq=%22city+reopens+old+rail+line%22&st=nyt.

Rogers, Heather. 2005. *Gone Tomorrow: The Hidden Life of Garbage.* New York: New Press.

Russell, Gordon. 2006. "Storm Debris Landfill is OK'd." *New Orleans Times Picayune,* April 14. [Online article; retrieved 6/24/07.] http://www.huongduong.org/giaoxu/tintuc/timespicayune%20041406.htm.

Stewart, Joshua. 2007. "Crofton Civic Group Dumps Landfill Deal." *The Capital,* April 17. [Online article; retrieved 4/17/07.] www.hometownannapolis.com.

Stroupe, Wayne. 2001. "Debris Removal Is Biggest Mission in New York." U.S. Army Corps of Engineers. November. [Online article; retrieved 6/24/07.] http://www.hq.usace.army.mil/cepa/pubs/nov01/story20.htm.

Tchobanoglous, George, and Frank Kreith. 2002. *Handbook of Solid Waste Management,* 2nd. ed. New York: McGraw-Hill.

U.S. Department of Agriculture. Forest Service. 2008. "Woody Biomass Utilization." [Online information; retrieved 6/14/08.] http://www.fs.fed.us/woodybiomass.

U.S. Department of Homeland Security (DHS). 2006. *Congressional Inquiry, Landfill Cost Issues Relating to Disposal of Debris in the City of New Orleans.* Report DD-07-03, December 15. http://www.dhs.gov/xoig/assets/auditrpts/OIG_DD-07-03_Dec06.pdf.

U.S. Environmental Protection Agency (EPA). 2006a. *Municipal Solid Waste Generation, Recycling, and Disposal in the United States: Facts and Figures for 2005.* Washington, DC: EPA.

U.S. Environmental Protection Agency (EPA). 2007. "Source Reduction and Reuse." [Online information; retrieved 7/11/07.] http://www.epa.gov/garbage/sourcred.htm.

U.S. Environmental Protection Agency (EPA). 2007a. "Brownfields Cleanup and Redevelopment: South Central Planning and Development Commission, LA." [Online article; retrieved 7/5/07.] www.epa.gov/swerosps/bf.

U.S. Environmental Protection Agency (EPA). 2007b. "National Priorities List (NPL)." [Online information; retrieved 7/5/07.] http://www.epa.gov/superfund/sites/npl/index.htm.

U.S. Environmental Protection Agency (EPA). 2007c. "The Solitron Microwave Site: From Groundwater Contamination to Economic Revitalization." [Online information; retrieved 7/5/07.] http://www.epa.gov/superfund/accomp/news/solitron.htm.

U.S. Environmental Protection Agency (EPA). 2007d. "Using Phytoremediation to Clean Up Sites." [Online information; retrieved 7/5/07.] http://www.epa.gov/superfund/phyto.htm.

U.S. Government Accountability Office (GAO). 2006. *Recycling: Additional Efforts Could Increase Municipal Recycling.* December. Washington, DC: GAO.

Walker, Rob. 2007. "Unconsumption." *New York Times,* January 7. [Online article; retrieved 7/11/07.] http://www.nytimes.com/2007/01/07/magazine/07wwln_consumed.t.html?scp=1&sq=%22unconsumption%22&st=nyt.

3

Worldwide Perspective

In July 2007, the Republic of Ireland raised its tax on plastic shopping bags, known as the plastax, from 15 to 22 cents after a study found that bag use had risen from 21 to 30 per capita in a single year. The Plastic Bag Environmental Levy—a charge that applies at the point of sale at about 19,000 retail outlets such as grocery stores, shopping centers, and gas stations—was first enacted in 2002 to discourage the use of plastic bags by consumers; an estimated 1.2 billion were handed out every year in Ireland. A national study published in 1999 presented the government with several options; researchers concluded that a tax on each plastic bag would best epitomize the "polluter pays" principle that targets end consumption, rather than trying to expand litter control measures by local authorities (Republic of Ireland 1999).

Almost immediately after the tax was implemented, per capita plastic bag use decreased from an estimated 328 bags to 21—along with a 95 percent decrease in plastic bag litter. Surveys showed that up to 90 percent of shoppers used life-long bags in 2003, compared with 36 percent in 1999. The levy brought in revenues of nearly US$100 million in its first five years, with proceeds going into the Environment Fund to support civic recycling centers, enforcement of Ireland's Waste Management Acts, waste awareness campaigns, the Green Schools Initiative, and efforts to manage used refrigerators and freezers. The 2007 increase was designed to bring per capita use back down to 2002 levels or lower (Republic of Ireland 2007).

The legacy of plastic bag litter is not limited to industrialized nations or large urban areas. In March 2002, Bangladesh banned polyethylene bags after researchers found they had blocked drainage systems and were a major cause of the floods that devastated the country in 1988 and 1998. South Africa, which calls plastic bags "the national flower" because they can be seen caught in bushes and along fences, began requiring manufacturers to make bags more durable and expensive in 2002, resulting in a 90 percent decrease in their use. Taiwan has adopted a law requiring restaurants, convenience stores, and supermarkets to charge customers for plastic bags and utensils, causing a 69 percent drop in plastic products, and in Australia, an estimated 90 percent of retailers are cooperating with the government's voluntary plan to limit plastic bags. In southwestern Great Britain, shopkeepers in the town of Modbury, population 1,500, agreed to stop giving out plastic bags to their customers and also declared a bag amnesty for those whose bags had piled up at home. In Paris, a law banning plastic bag use went into effect in 2007, to be followed in 2010 by a nationwide ban.

Like plastic bag litter, waste and its management is a problem that affects every society, from the most developed countries to those considered part of the Third and Fourth worlds. The issue is complicated because waste involves those countries that produce waste, those that recover and reuse waste, and those that trade waste among themselves. Another factor is the wide variation in waste disposal methods, depending on a country's resources. In the United Kingdom, for instance, a lack of readily available space for landfills makes incineration one of the most used methods for eliminating waste, a situation that is similar to that of Japan. In Yokohama, the 3.5 million residents sort their trash based on a 27-page booklet that details instructions for 518 items. Garbage must be sorted into 10 categories as a way to reduce incinerated trash by 30 percent over the next five years in a country where space for landfills is almost impossible to find (Onishi 2005). In Australia, the population is less dense and more dispersed, making landfills the most common strategy for waste disposal. Other countries, especially those in Scandinavia, employ waste-to-energy both as a waste disposal method and to produce needed heat or power for local communities. Just outside Toronto, for example, a subsidiary of Algonquin Power Income Fund burns 500 tons of waste per day and generates 15 megawatts of electricity.

This chapter outlines the primary problems facing waste managers from a worldwide perspective and the solutions that are being attempted to solve them. It does not address the waste issues of every country, but it does provide a snapshot of what is being done from one point on the waste continuum to another.

Global Waste Management Efforts

Africa

Progress toward environmental sustainability in Africa has been slow because of the need to prioritize basic human needs for housing, food, and security. But in some African countries, environmental consciousness is gradually being raised as the standard of living increases and the amount of pollution and waste created mounts. The "poverty-environment-growth nexus" is becoming better understood as the paradigm shifts from taking action after waste is generated to finding ways to prevent pollution through waste minimization (Ekelund and Nystrom 2007).

Waste management in most of Africa centers on two problems: the lack of accurate data on solid waste generation and characterization and a corresponding lack of information about waste collection, processing, and disposal. Thus, even the most recent studies are unable fully to determine how much progress is being made. A July 2002 report by the African Development Bank found, for instance, that no country has specific waste management legislation, although some statutes are being drafted, and virtually no integrated waste management programs are being implemented. The lack of information about what is being done officially is, to some extent, supplemented by research on the informal trash economy (scavenging and privatization). But overall, education and communication channels between the government and civil society are considered inefficient and inadequate. "A lack of a right to know, secrecy and misinformation has also been major contributory factors to poor waste management in many African countries" (African Development Bank 2002).

Case studies of four cities with a population of 1 million or more show major differences in efficiency and practices in Africa. In Cairo, for instance, the government has established modern composting plants that have the potential to serve as a model for other African countries because of their modern design, low cost, high efficiency, and satisfactory operation records. But engineered landfills are only in the planning stages, and a need exists for a long-term waste management plan and a focused public education campaign. One serious problem in Nairobi, in contrast, is the lack of regular collection of municipal solid waste, with only 20 to 30 percent of waste collected. Trash pickup vehicles are often unavailable because of mechanical problems and a lack of spare parts, so some portions of the city's streets are swept manually; mechanical street sweeping is unavailable. Some trash is picked up by private companies, which transport it to an unfenced dumping area that is open to scavenging and cattle growers. Accra, Ghana, operates a composting plant (although it does not work at full capacity) and a regional dump. Based on strategic objectives outlined in a 1999 sanitation policy, the government of Accra hopes to have all solid waste generated in urban areas collected and adequately disposed of by 2020. In Cape Town, South Africa, more than 95 percent of domestic, trade, and hazardous waste is landfilled, the most widely used and cheapest option in the country. South Africa has transfer stations, privately owned landfills, recycling programs (about 7 percent of waste is recycled), and plans for an integrated waste management approach (African Development Bank 2002).

South Africa held its first National Waste Summit in September 2001, bringing together representatives from government, business, and nongovernmental organizations (NGOs) to find ways to reduce waste. The resulting Polokwane Declaration called for a vision of zero waste by 2022. One of the strategies considered by officials was composting, because it has the most considerable environmental benefit. Municipalities can reduce the amount of waste going to landfills, which in turn reduces the environmental impact of the remaining waste by limiting the amount of organic waste that can lead to methane gas production, leachate, and the risk of fire. Because compost can also be used as a fertilizer, it has other benefits to local residents.

But in most major African cities, informal waste recycling is still an important, if unrecognized, aspect of waste management. More than 50,000 *zabbaleen*—an informal network of trash scavengers—recover and recycle between 70 and 80 percent of all collected recyclables in Cairo, for instance. Swine operations, in which pigs are fed garbage, produce food and a secondary income source for their owners, who can sell the organic waste. Most experts believe this scavenging system is efficient and innovative, but waste picking is seldom recognized by the central government, and the health problems and poor working conditions that result are often ignored (African Development Bank 2002).

A 2007 United Nations (UN) Environment Programme study of poor countries in Africa found that a lack of zoning and inadequate enforcement of existing laws often lead to the growth of slums surrounding dumps in cities. The 30-acre Dandora dump, for instance, located five miles from Nairobi, pollutes the entire region with toxic smoke from burning trash. An estimated 2,000 tons of garbage, including industrial and medical waste, are taken to the dump site every day. Residents wash their clothes and bathe in the Nairobi River adjacent to the dump; downstream residents are exposed to hazards from runoff and leaching. Medical tests and soil samples found that children in the area had high levels of lead in their systems, and many also suffer from respiratory problems from scavenging in the odorous trash for items they can sell. Asthma, anemia, and skin infections are common; officials said "the actual results are even more shocking than we had imagined at the outset" (Muhumed 2007).

Asia

In July 2000, environmental activists from 12 Asia-Pacific nations met in Bangkok, Thailand, to found a regional alliance, Waste Not Asia, to oppose waste incineration and promote ecological methods of waste management. The group's goal is to achieve a sustainable society with zero waste through programs that emphasize clean production, materials recovery, and community-based solutions. The anti-incineration emphasis is the result of allegations that Japan and Europe have forced toxic technologies on Asian nations, "which seeks to make Asia the toilet bowl of the industrialized world. We are here to let the world know we no longer intend to become the dumping ground for their discarded technologies," according to one group member (Grass-Roots Recycling Network 2000). Subsequent meetings were held in Taipei, Taiwan, in 2001; in Penang, Malaysia, in 2003; in Seoul, Korea, in 2004; and in Kerala, India, in 2006.

Particular attention has been focused on China, whose environmental record has not been consistent with its industrial growth, and concerns about how the world's most populous nation is dealing with its own waste. China was one of the first countries to join the Basel Convention on the Control of Transboundary

Movements of Hazardous Wastes and Their Disposal, described in more detail later in this chapter. In 1995, the National People's Congress adopted the Law on the Prevention and Control of Environmental Pollution by Solid Wastes of the People's Republic of China. The goal of the legislative action was to minimize hazardous wastes to the greatest extent possible, setting down management systems such as a system of waste registration, a hazardous waste permitting system, and transport monitoring. In 1998, China's State Environmental Protection Administration, responsible for implementing the Basel Convention, issued a national catalogue of hazardous waste comprising 47 types, listing their sources and the hazardous substances they contain. Other regulations have been developed to implement a permit system for waste imports, although officials note that China itself produces some hazardous wastes for which it has no treatment facilities, necessitating that those wastes be exported to other countries for disposal (Zhenhua 2007).

China's stance on the disposal of radioactive waste is important because the government plans to boost its current nuclear power generation capacity by 20 times by the end of 2030. Ten nuclear reactors were generating power in 2007; those plants produce about 8 million kilowatts of power. By 2020, the governments wants to increase generation to 40 million kilowatts, and between 120 million and 160 million kilowatts 10 years after that. To achieve these goals, China would have to build in excess of 100 nuclear reactors, each capable of generating 1 million kilowatts over 20 years. If accomplished, these efforts would make China the world's largest generator of nuclear power, surpassing Japan, France, and the United States. Although these plants would be new, and it could be decades before they would be decommissioned, the amount of waste they would create is of concern to other countries in the region.

India's waste management issues are much different from those of neighboring China. Because of cultural and economic conditions, only about 60 percent of solid waste is collected nationwide, with the remaining 40 percent just dumped on the waysides and in urban streets. The country has no comprehensive resource management plan and no laboratory to test for the presence of chemicals such as dioxins that are produced from waste incineration. In 2000, the Indian Supreme Court ruled that municipal waste agencies should set up waste processing and disposal facilities by December 2001 and identify future landfill sites by December 2002. All local governments were expected to have door-to-door collection systems and have segregated waste programs for recyclables and organic waste in place by December 2003. But compliance with the Municipal Solid Waste Management Rules (MSWMR) 2000 has been only partially successful, studies show. Some municipal governments still have not developed waste services at all, while others depend on NGOs working independently to provide collection of waste. Most programs exist on a very small scale despite the size of the population, as is the case in Mumbai. Companies attempting to build capacity for recycling plastic bottles have had to import bottles from the United States in order to make their operations financially profitable (Sule 2004).

Like many poor countries, India is also home to thousands of *ragpickers*—the term used to describe trash scavengers—who are trying to become more organized and to be integrated into the nation's developing waste management programs. Some have created trade unions by quantifying their contribution to solid waste management and using collective bargaining power to create state recognition of their efforts. Some groups have been hired by contractors, although many companies discriminate against women. The creation of the MSWMR has resulted in the elimination of jobs for many, leaving ragpickers destitute (Sule 2005). Some critics have argued that municipal governments have ignored the plight of the ragpickers and of residents in the poorest areas of India. In June 2007, for instance, 25 trucks carrying garbage from a city corporation rolled into a small village with a police escort to dump waste at a 37-acre site where a waste treatment plant was proposed. Although the local residents tried to stop the trucks, they were beaten by police; more than 200 people had to flee the area because they became sick from the garbage stench, and some had to be hospitalized. The corporation had secured a permit to operate a waste facility, but not for open dumping that contaminated the soil and water. A member of an independent fact-finding committee called the dumping, which continued for five days, "a blatant human rights violation" (Suchitra and Venugopal 2007). Although some ambivalence about building on the site had been encountered, environmental and health concerns have gained increased attention, since the facility is to be built near a water source used by both residents and a host of industrial users (Basheer

2007). The city corporation did not have an alternative plan for dealing with the waste from a densely populated area after years of just dumping it in open areas or along roadsides.

While some might perceive India to be decades behind the rest of the world in its waste management practices, success stories can be found. In Suryapet, a town in the Nalgonda district in Andhra Pradesh, the city's inhabitants generate more than 48 tons of garbage per year. Each resident is given two bins—a red one for dry garbage and a green one for wet garbage (food, flowers, etc.)—which are collected in a door-to-door system each morning. The city has no need for a landfill because all waste is either composted or recycled (Krishna 2006). But these programs are in the minority, evidence of the difficulties of trying to create a waste management infrastructure in areas that are already experiencing industrial or residential growth.

At the other end of the scale is Japan, which has a long history of municipal waste management. Rapid post–World War II industrialization resulted in a flurry of environmental legislation during the late 1960s and early 1970s, paralleling activity in the United States. The 1967 Basic Law for Pollution Control and the 1970 Pollution Diet forced officials to balance out the conditions associated with a densely populated country, with typical urban problems, such as waste, and the government's desire to build a strong economy. Responsibility for waste management is decentralized and fragmented among numerous centralized agencies and mostly privatized or in public-private partnerships. About half of all hazardous waste is dealt with on-site by manufacturers, so disposal firms tend to be small and owned by domestic companies. Landfills are scarce; about half of the country's toxic wastes are disposed of through incineration—the highest rate in the world (O'Neill 2000).

Its more serious problem is radioactive waste. Japan began its commercial nuclear power plant operations in 1966, and now nuclear power accounts for about one-third of Japan's electricity production. Low-level waste generated by nuclear power plants has been buried at the Rokkasho-mura facility since 1992, and spent nuclear fuel is routinely stored on-site at reactors in pools until it is sent to France or the United Kingdom for reprocessing and then returned to Japan. High-level radioactive waste is vitrified (converted to a solid form) and stored on-site for cooling, with the intention that it will eventually be placed in some sort of deep geologic disposal facility as Japan's first option for long-term storage. Dry storage is also being developed at the Rokkasho-mura site. Facility development falls under the auspices of the Nuclear Waste Management Organization, which plans to begin construction in the 2030s.

Over the last 25 years, Japan's government has focused its attention on the technical capacity of its research and development programs to store high-level radioactive waste, while the public debate has centered around concerns over siting of a facility. Japan's active tectonic setting and complex geology make the issue problematic, as was demonstrated when a 6.8-magnitude earthquake struck the world's largest nuclear power plant at Kashiwazaki-Kariwa in July 2007. The plant, on the coast of Japan about 135 miles north of Tokyo, was ordered closed after it was reported that the damage to the facility was much greater than was originally announced. Japanese officials have prided themselves on the transparency of their nuclear program, but the incident did little to calm public fears about how an earthquake might damage an underground depository.

European Union

Waste management is not usually front-page news in Europe, but when it does make headlines, it is for good reason. Naples, Italy, suffered what could only be described as a garbage crisis in 2004, when uncollected trash began piling up on the city's streets and then showed up throughout the picturesque countryside, where it was dumped. The scene was repeated in May 2007, when alleyways became blocked by piles of trash and plastic bags full of debris began showing up on beaches on the Amalfi Coast and the islands of Capri and Ischia. Even though residents protested, little was done to alleviate the growing mounds of trash that provided homes for rats, cockroaches, and swarms of flies. Officials in the tourist city said that tourists were encountering waste-filled streets as they attempted to see ancient sites like Pompeii and Mount Vesuvius.

The problem was more than just visual, however. City health inspectors began testing the air quality around some of the largest piles of garbage, and many of the city's school children were given masks because of the stench. Local officials urged residents and tourists to avoid the large packs of stray dogs seen scavenging through the trash heaps, and banks, hotels, and restaurants faced closure if the air quality risk continued. The mayor of Naples estimated that more than 3,000 tons of garbage choked city streets by the start of the summer tourist season, with no resolution in sight. Italy's president called the situation "tragic" and noted, "This is as bad an emergency as Naples has ever had" (Nadeau 2007).

The Naples garbage problem is a result of two forces: infighting among the organized crime syndicates of the Italian Mafia that controls the waste management industry and the "not in my backyard" attitude of residents opposed to the siting of four proposed landfills. Italy produces an estimated 80 million tons of waste a year, of which 35 million tons are estimated to be handled by criminal organizations. More recently, the estimated 158 families that make up the "eco-mafia" moved into the collection and handling of toxic waste material. Trafficking in toxic waste did not become a crime until 2001, years after waste dumps were being found throughout the country. According to one organization, nearly 12 tons of toxic garbage "disappeared" in 2001 in Italy. Sometimes, the waste is camouflaged as loads of fertilizer or mixed with cement or asphalt. In 2002, the police discovered a clandestine business engaged in waste collection, transport, and elimination in which six municipal officials and two local business executives were involved—a common way for organized crime to operate with virtual impunity (Colombo 2003).

One proposed site for a new Italian waste facility is near a World Wildlife Federation Park and another is adjacent to a rail line used by tourists visiting the Amalfi Coast. Local residents decided to take matters into their own hands by starting fires to burn the trash, and officials had to place signs on dumpsters asking people not to burn rubbish because of concerns about releases of airborne toxins. The combination of summer heat, decomposing garbage, and foul odors even led to a threatened resignation by Italy's "garbage czar." Trash was transformed from a product of urban existence to an ecological crisis that made international headlines.

Italy is not unusual. An estimated 1.3 billion tons of waste are created each year in the European Union (EU), of which 40 million tons are believed to be hazardous. The European Environment Agency estimates that each person in the EU generates about 3.5 tons of solid waste and 700 million tons of agricultural waste each year. The Organisation for Economic Co-operation and Development contends that by 2020, the EU could be generating 45 percent more waste than in 1995 (European Commission 2007).

The EU's waste management approach is based on three principles: waste prevention, recycling and reuse, and improving final disposition and monitoring. These principles are codified in Directive 2006/12/EC of the European Parliament and of the Council of 5 April 2006 on Waste. The framework agreements require member states to prohibit the abandonment, dumping, or uncontrolled disposal of waste. The measures also call for cooperation among member states with a view to establishing an integrated network of disposal facilities to enable the EU to become self-sufficient in waste disposal. The EU countries also have agreed to support the polluter-pays principle, requiring the cost of disposing of waste to be borne by the entity that creates the waste or has it handled by a waste collector (Europa 2007).

The standards for dealing with hazardous waste vary considerably from one country to another within the EU, however. At one end of the spectrum is the Netherlands, where the government's regulations concerning toxic soil contamination are strictly enforced, regardless of the financial burdens involved in cleaning up the waste. The Interim Soil Cleanup Act was enacted in 1981 as a response to contamination in the city of Lekkerkerk, a situation whose circumstances were similar to Love Canal in the United States. This act was followed by legislation that made the national government responsible only for major instances of waste remediation. Because the number of cleanups turned out to be much higher and the cost more expensive than had been originally anticipated, the government established a tax on companies that treat, store, or dispose of chemical waste, an action that had the unintended consequence of providing an incentive for firms to export hazardous waste or dump it illegally. The tax was replaced by a fuels tax in 1988 (Page 1997, 150).

In the United Kingdom, in contrast, land contaminated by waste soil is less likely to be seen as a hazard, and the government appears to have taken a less cautious approach. The 1995 Environment Act sought to establish an optimum balance between environmental protection and the economy, which is more similar to the approach of the United States. The standard that was established for contaminated land was whether significant harm was being caused or a significant possibility existed of harm being caused, a threshold that removes a large proportion of former industrial sites from the contaminated category (Cairney and Hobson 1998).

Along with the variations in how hazardous waste is categorized are the differences in approach to liability. Although the EU countries have agreed to support the polluter-pays concept of assigning the responsibility for remediation to the party that caused the contamination, similar to the method used in the United States, the EU's approach has one major difference: If the government is unable to find a polluter capable of paying for the cleanup, the European countries generally use government funds to pay; in the United States, extreme legal powers are used to force private parties to pay for the damage, even when the liability is only for potential responsibility or retroactive responsibility.

In Western Europe, major efforts are now being made to return formerly contaminated sites, or brownfields, to productive use. One major reason for this effort is the fact that these countries are more densely populated than the United States and fewer alternative sites are available. In addition, in most European countries, more strict controls on land use are in place, and the government can broker deals and assist companies in redevelopment. The use of place-based subsidies allows governments to provide incentives for private parties to develop previously contaminated sites. The United Kingdom has backed into environmental cleanup of wastes, doing so not because of environmental concerns but rather because of concerns connected to the economic revitalization of blighted urban areas (Page 1997).

In Central and Eastern Europe, where the hazardous waste sites are considered the worst in the world, civilian and military operations caused spills, leaks, and improper disposal of almost every type of chemical and contaminant known to exist. Former industrial facilities, military bases, and energy producers were not regulated by a central authority, and little concern was given to environmental degradation. Even with the gradual transition to market economies and democratic governments in the countries of this region, many of the sites in Central and Eastern Europe are not being cleaned up because of a lack of technology and cash. Countries such as Hungary, for instance, have a limited budget for remediation, thus, much of the effort is focused on containing the spread of contamination rather than cleaning it up (Page 1997). Some of these countries face the additional problem of being the disposal site for hazardous waste shipped there from abroad. Prior to 1990, the bulk of waste from Germany was exported to what was then East Germany; Germany now exports its waste to France and the United Kingdom, consistent with its position as a net exporter of hazardous waste even though it offers high-quality, state-of-the-art facilities to treat toxic waste (O'Neill 2000).

In the former Soviet Union, additional concerns have arisen about radioactive waste. In 1993, for instance, Western observers had already estimated that radioactive materials had been stored in 320 cities and 1,548 other locations within Russia, along with more than 200 tons of radioactive waste stored at research centers in Moscow (Feshbach 1995). Since that time, new discoveries continue to be made of sites where radioactive materials were buried in subsoils or in leaky containers that may be contaminating groundwater. Much of the country is still off limits to foreign monitors and researchers, so no accurate accounting is available of what may lie undiscovered in dumps there.

Radioactive waste has been a subject of great public interest in Europe because nuclear power is a significant source of energy in many countries. Only five EU states have active nuclear power programs (Finland, France, Spain, Sweden, and the United Kingdom); two large reprocessing facilities, one in Sellafield in the United Kingdom and the other in La Hague, France, are operating. All high-level waste is stored in surface or near-surface facilities because no permanent facility has been developed in any of the five nations. The issue is complicated because no new nuclear reactors were built in Europe for 15 years, and after such a lengthy lapse in development, technological expertise is lacking.

In Finland, where about one-third of the energy comes from nuclear generation, waste was routinely shipped to the Soviet Union until 1996, when the Finnish government ruled that its nuclear waste must be handled within Finland itself. The Finnish parliament approved the siting of an underground laboratory for a deep geological disposal site in 2002, and construction on the site began in 2004. It is expected to be operational by 2020. In France, nuclear energy is the main source of electrical power, comprising about 75 percent of the nation's energy production through 58 reactors. Officials have been studying deep disposal, but enormous public opposition to investigating potential sites for storage has been encountered. Spain, which has an abundance of reasonably priced uranium, had planned to develop its own reprocessing facilities for spent nuclear fuel but canceled its plans to do so in 1983. Its present policy for high-level nuclear waste is continued interim storage; in 1999, the Spanish government approved a plan that would delay any decision on what to do about deep geologic disposal until after 2010. The process of choosing a site for a deep repository in Sweden is now underway, with a final decision due in 2010 and shipments of radioactive waste expected to begin in 2015. The United Kingdom has conducted numerous studies on what to do about the 10,000 tons of radioactive waste currently stored there, including the 250,000 tons of waste that is expected to need some type of storage solution as existing nuclear power plants reach the end of their working lives. Various options have been considered over the years, along with a proposal that a strong, independent, authoritative body be established to make recommendations, and that a lengthy public consultation process take place.

Germany, which is backing away from nuclear power by announcing it will shut down all nuclear reactors in the country by 2020, identified a massive underground salt chamber for storage in 1977, but research there was halted in 2000 because of political challenges. Several other underground sites for the indefinite storage of highly radioactive waste were identified in 2007, but antinuclear activists are unconvinced that the scientific and technological issues have been solved (Hawley 2007).

The European Commission adopted a proposal dealing with the management of spent nuclear fuel and radioactive waste as part of its five-document "nuclear package" in 2003. The proposal is part of the attempts to harmonize standards and practices within the countries of the EU. An estimated 40,000 cubic meters of radioactive waste are produced each year in the EU, the majority of which comes from nuclear power generation. However, no European Union–wide policy is in place for storing or disposing of the waste, so each country is responsible for developing its own policies and solutions. But even if a site were to be approved today, at least a 20-year-long wait would ensue before a facility is ready to accept shipments. As one official notes, "That is simply a different time scale than we are used to in political life. There is not a very big motivation for politicians to make a decision" (Hawley 2007).

Latin America

Considerable differences can be seen in how waste management services are provided in Latin America, where more than 70 percent of the people live in urban areas. According to the U.S. Agency for International Development (USAID), the quantity of waste generated per person in Latin America is increasing, and the composition of household and business waste has shifted from being almost entirely degradable to being considerably less so. There are also variations in how well solid waste is managed. In the cities and maquiladoras (industrial facilities) along the U.S. border with Mexico, the waste materials from the factories are routinely collected by companies with formal contracts that pick up the waste for recycling. In large cities such as Caracas, Venezuela; Brasilia, Brazil; La Paz, Bolivia; and Medellin, Colombia, collection rates are as high as 90 percent, with rates in medium-size cities at about 57 percent, and in smaller cities and towns, coverage is limited. While solid waste is collected at high rates in cities, rapid expansion has created underserved populated areas where the situation is more daunting. This is especially true in tropical areas (USAID 2003).

Collection is just one phase of waste management, however, and the disposal practices are not as well developed. Very little of Latin America's solid waste goes into sanitary landfills; most is deposited in open dumps or landfills where groundwater is easily contaminated by leachate. Almost no source separation takes place at these sites, and treatment of organic waste is nearly nonexistent. Although more than a dozen large-scale composting facilities have been built, all or most are not operational. One successful element of waste management deals with recycling, where recovery is higher in Latin America than in many industrialized countries; in Brazil, for instance, 87 percent of all aluminum cans are recycled.

The political and institutional arrangements for waste management range from highly structured to almost nonexistent. Some cities have formal contracts with companies to collect and dispose of trash, and others utilize concession-type schemes. In still other communities, garbage pickup and disposal are viewed as municipal services and are provided by city employees. In Guatemala, all service is contracted privately. Most urban areas also utilize some form of informal scavenging. In Buenos Aires, Argentina, the city government passed a law in 2003 that requires trash pickers to register with authorities and to receive an official license to scavenge. The scrap workers (or *cartoneros,* who prefer to be called *recuperadores*) are somewhat legitimized and have been integrated into the city's sanitation system. Those who register may have access to a health plan, receive vaccinations, and benefit from other government services. In Paraguay, a loan and grant program for the families who scavenge dumps at the capital city of

Asunción is designed to raise their income and reduce poverty levels. The NGO Alter Vida assists the *gancheros* (another name for scrap workers) by promoting sustainability through the donation of trash collection trucks and the building of a collection and sorting facility to give the scavengers a healthier and safer place to work (Constance 2005).

Special wastes (hazardous, medical, industrial) are inadequately or improperly handled throughout Latin America. Waste collected from hospitals or other medical facilities, including pathogenic materials, is routinely deposited along with household waste in dumps, or is burned. Some industrial waste is deposited along the roadside or dumped illegally, with exceptions in countries like Brazil, which has licensed incineration facilities. Chile has a hazardous waste and treatment facility in Santiago, and Peru, Colombia, and Ecuador have designated segregated areas for depositing hazardous and medical waste. But the reality is that the waste must be collected and transported to the landfill for disposal, steps that often are not undertaken in favor of illegal dumping (USAID 2003).

Middle East

The political and civil unrest that has typified the Middle East for decades has made basic services such as waste management a symptom of government collapse. Piles of trash are a frequent sight in the region, as refuse pickup is one of the lower-prioritized services provided. In some densely populated cities, such as Baghdad, trash collectors' lives are at stake because insurgents frequently use piles of garbage in which to hide bombs. Most of the municipal workers who have been killed in the city are trash collectors who make only a few dollars per day. Before the fall of Iraq's dictator, Saddam Hussein, in 2003, 1,200 trash collection trucks were operating in Iraq; by 2007, only a third that many were operating in a city of 6 million people. The mounds of refuse are seldom collected, even in affluent neighborhoods, because workers fear being out in the open and lack security. The buildup of trash has led to public health concerns over flies, mosquitoes, and odors from poor sanitation. Residents who once disposed of their trash in municipal bins now dump it alongside roadways, where it may never be hauled away (Luo 2006).

The exception in this region is Israel. In 2005, each person generated an average of 3.37 pounds of waste per day, translating into more than 6 million tons of solid waste (domestic, commercial, and industrial). Israel's population has been increasing at about 2 percent per year, but solid waste volume has grown about 5 percent annually, because of the rising standard of living. Israel has limited land resources for dealing with solid waste and has implemented an integrated waste management plan similar to that of the United States, featuring source reduction, reuse, recycling, energy recovery, and landfilling. In 2004, the Israeli recycling rate was about 20 percent, up from 3 percent 10 years earlier. The target is a 50 percent rate by 2010. To reach this target, policy makers have adopted several important steps, from closing old dumps and replacing them with alternative, environment-friendly sites farther from population centers to building new sorting plants as an alternative to landfilling (Israel Ministry of the Environment 2007).

Israel enacted its Hazardous Substances Law in 1993, which encompasses a "cradle to grave" system that requires companies to track and monitor substances from the point where they are created through waste disposal. This system includes licenses, regulations, and supervision of some aspects of production, use, handling, marketing, transport, import, and export. In 1994, the government promulgated hazardous waste regulations as a legal basis for implementing the Basel Convention discussed later in this chapter. Another series of regulations was adopted in 2002, dealing with radioactive waste by setting prohibitions, obligations, and limitations on the disposal of these wastes.

More recently, a private Israeli company has developed a process using plasma gasification melting technology that combines high temperatures and low-radioactive energy to transform radioactive, hazardous, and municipal waste into inert by-products such as glass and clean energy. The process was developed in conjunction with Russian researchers by Environmental Energy Resources (EER) in 2007, and its supporters note that it produces no surface water, groundwater, or soil pollution. The solid material that is produced by the process can be used by the construction industry for tiles and blocks. EER has used the process, which costs about $3,000 per ton in comparison with $30,000 per ton for treating and burying low-level radioactive waste, in the United States and Ukraine. The company was approached by officials looking for a way to treat low-level waste resulting from the 1986 meltdown at the Chernobyl nuclear facility. The process operates turbines to generate electricity, about a third of which can then be sold (Kloosterman 2007).

Oceania and Micronesia

This Pacific Island region represents a mix of both wealthy and highly industrialized countries and poor, developing nations. For instance, Australia, like Japan, has a well-integrated municipal waste management program. The system is privately run, with involvement by large multinational companies such as Waste Management International. A major controversy erupted in the late 1980s over a national waste disposal facility and a high-temperature incinerator that was to be publicly owned but privately managed. The plans for the incinerator were opposed by NGOs such as Greenpeace, and the proposal was dropped in 1997.

Australia's two primary waste-related problems are hazardous waste and the issue of low- to medium-level nuclear waste storage. Australia's federal constitution, which was enacted in 1900, does not mention environmental issues, and until the early 1970s, environmental problems were handled by the six states and two commonwealth territories, rather than by the central government. States have jurisdiction over environmental problems within their boundaries, and until recently, few institutions or mechanisms were in place to deal with transboundary issues. Hazardous waste is generated and disposed of in only two states, with the cities of Sydney and Melbourne accounting for 70 percent of Australia's industrial waste production. In 1992, the government announced a two-year moratorium on the export of hazardous waste to less developed countries, and the extension became a total ban in 1996 (O'Neill 2000).

The bulk of Australia's radioactive waste comes from two research reactors and facilities where individuals are treated for illnesses and includes items such as contaminated clothing, gloves, glassware, and soil. If proposals to store the waste were accepted, spent fuel from the reactors that previously was sent overseas would be brought back and stored in aboveground containers. Commonwealth officials have considered several locations for a facility to dispose of this waste, including three government-owned defense sites. The most controversial proposal, however, comes from an offer from Aboriginal landowners in the Northern Territory who have agreed to allow the facility to be located on their land for 200 years under a long-term lease. In exchange, the government would give the land back to the traditional owners after it was declared safe, along with $10 million to be managed by a charitable trust. Some indigenous leaders believe the money will go a long way toward improving living conditions and educational opportunities for clan members, while others feel the dump would poison the land, which was given back to the clan in 1995 after bitter disputes and court fights. One elder of the 70-member Ngapa clan said that the deal would "create a future for our children with education, jobs, and funds for our outstation and transport." Another clan member said the facility would "change our dreamings," while a member of the Australian Conservation Foundation said the facility was not selected on a scientific basis, and turning the sparsely vegetated land into a dump would be "environmentally irresponsible and socially divisive." A government official contended, "This potential facility could compromise the social, cultural, and traditional ties of Aboriginal people to their country" (Murdoch and Afianos 2007).

If the nominated site is approved, Australia's nuclear industry would be the ultimate beneficiary. The prime minister has envisioned up to 25 nuclear generating stations within the next 20 years, but the primary barrier has been the refusal of state governments to allow either the facilities or the waste site to be built on their land. Having the indigenous people of the Northern Territory offer their land allows the federal government to move forward despite state government objections. The facility is scheduled to begin operation in 2011 (*Earthtimes* 2007).

Throughout the rest of Micronesia, the situation varies from one island group to another. Saipan, which became part of the U.S. Commonwealth of the Northern Mariana Islands in 1978, has a population of about 80,000 and falls under the jurisdiction of the U.S. Environmental Protection Agency (EPA). The island is remote and home to a thriving garment industry that is responsible for about a third of Saipan's solid waste stream. Because Saipan is subject to U.S. regulations and is developing its tourist industry, it has implemented one of the most technologically simple but efficient waste management strategies in Micronesia. Officials closed an old military dump along a lagoon that had been used as a landfill and built a $9.4 million solid waste facility to replace it. The island also has a refuse transfer station, solid waste diversion and recycling programs, and a public outreach program that is considered a model for the rest of the region (Hiney and Hawley 2005). But Saipan's successes could not have come without funding (80 percent comes from the U.S. Department of the Interior) and the regulatory force of the EPA.

Most of the other islands in Micronesia face major pollution hazards and lack regular garbage collection systems and landfills. Dumps are overflowing, and shipping garbage is not a reasonable option. In Guam, the 50-year-old municipal dump is unlined and uncontrolled, and trash collection trucks have frequent breakdowns. Residents have complained of having to wait four weeks to have their trash picked up, resulting in illegal dumping of trash in the woods or along beaches. But the government of Guam does not have the financial resources to support a new municipal dump, and change is unlikely unless it comes from the private sector.

International Waste Management Regimes

Although each country is responsible for managing its own waste, several attempts have been made to find a global solution to problems that cross sovereign borders. Some of these efforts take the form of international agreements, or regimes, while others are activities coordinated by NGOs. One of the factors that makes the international control of waste difficult is that every country has its own definitions about the universe of waste, so statistics for comparison are difficult to come by. There is no doubt that the volume of waste is increasing, but the attention has not been sufficient to put pressure on international bodies to develop significant new controls to better manage waste.

Hazardous Waste Trading

The issue of hazardous waste trading gained salience in the late 1980s as the industrialized countries developed their own environmental protection legislation and regulations. The result was a tremendous increase in the cost of hazardous waste disposal, and "toxic traders" began looking for cheaper ways of disposing of their wastes. The primary destination for hazardous waste exports was Eastern Europe and other developing countries. Media coverage and action by NGOs, especially Greenpeace, led to a series of international and regional agreements designed to prohibit hazardous waste trading and, later, to efforts to reduce hazardous waste at the source of production.

Basel Convention

After two years of negotiations involving 116 countries, the international regime known as the Basel Convention on the Control of Transboundary Movements of Hazardous Wastes and Their Disposal was adopted in Basel, Switzerland, on March 22, 1989, as a response to the surge in international waste trafficking that began in the late 1980s. The convention regulates the movement of transboundary waste and requires the parties to the agreement to ensure the waste is managed and disposed of in an environmentally sound manner. The convention entered into force on May 5, 1992, with 170 nations currently listed as parties, from Albania to Uruguay, although the United States has not yet ratified the convention. Because the United States is a nonparty, many countries that are parties to the convention are prohibited from exporting their waste to or importing waste from the United States. The member states of the European Union are subject to the ban because of agreements related to international trade, which actually increases the number of implementing countries. In 1995, the Basel Convention was amended to ban the shipment of wastes destined for disposal or recovery from developed to less developed countries. The Basel Ban Amendment is extremely controversial and has not yet entered into force.

Under the initial agreement, wastes are defined as "substances or objects which are disposed of or are intended to be disposed of or are required to be disposed of by the provisions of national law." The convention also considers "disposal" as deposit into or onto land, land treatment (such as biodegradation), deep injection, incineration on land or at sea, permanent storage, repackaging, recycling, and recovery. "Hazardous" wastes are those stemming from particular manufacturing processes and hazardous constituents of wastes such as copper compounds, lead, and organic solvents, or wastes that are defined as or considered to be hazardous under the domestic legislation of the country of export, import, or transit. Precise definitions of what are considered hazardous waste are found in the various annexes to the agreement.

The convention also places conditions on the export and import of waste, as well as strict notice, consent, and tracking requirements for the transboundary movement of wastes. Each party is required to develop a national reporting system, which is reviewed by the secretariat to the convention, located in Geneva.

The parties are prohibited from importing or exporting waste if there is reason to believe the wastes would not be managed in an "environmentally sound manner," defined as "taking all practicable steps to ensure that hazardous wastes or other wastes are managed in a manner which will protect human health and the environment against the adverse effects which may result from such wastes," which some critics believe is open to considerable interpretation (Hunter, Salzman, and Zaelke 2002). This definition makes the agreement highly subjective and difficult to enforce.

Although any illegal trafficking in hazardous waste is considered a criminal offense, one of the weaknesses of the Basel agreement is the lack of any enforcement capacity. The regime includes an extensive definition of what is considered illegal waste trade and requires the initiating party to take waste back if it is found to be in violation of the agreement or if the receiving party is unable to manage the waste. Any efforts to take action must be initiated by the parties themselves through their own legislative, administrative, or judicial institutions and processes.

The convention has also been criticized because, despite an initial burst of momentum when the agreement was first adopted in 1989, ratification was considerably less enthusiastic. Of the six largest waste-producing countries—the United States, Canada, the United Kingdom, Australia, Japan, and Germany—only Canada and Australia ratified the convention before the end of 1992; Japan ratified in late 1993; and Germany and the United Kingdom signed on in late 1994. Developing countries also delayed the ratification efforts, especially the African states that sought their own, stronger agreement (Clapp 2001).

One of the most recent controversies surrounding the Basel Convention deals with Japan and its efforts to use bilateral trade agreements that eliminate tariffs on hazardous waste. The NGO Basel Action Network (BAN) submitted a Basel Non-Compliance Report (a type of formal complaint) against Japan in March 2007 alleging that the country was violating international law. In a letter to the UN Environment Programme, BAN accused Japan of "throwing its economic weight around among developing nations to force them to take Japanese hazardous waste" (Environmental News Service 2007). The group cited three partnership agreements with Singapore, Malaysia, and the Philippines that would allow Japan to send its hazardous waste abroad by providing the three nations with financial and economic assistance. Characterized as "waste colonialism" by detractors, the economic partnerships are also being planned for Thailand, Indonesia, India, Australia, Vietnam, and Chile. A representative of the Global Alliance for Incinerator Alternatives noted, "It is both an affront to international law and morally repugnant for a rich country such as Japan to bully poorer nations with its money into accepting its toxic effluent" (Environmental News Service 2007).

One of the NGOs that has been actively seeking solutions is the International Solid Waste Association (ISWA), based in Copenhagen, Denmark. The organization includes members in 32 countries and works in cooperation with the UN Environment Programme, the World Bank, and the European Union. Many of the group's objectives are based on the 2002 Johannesburg World Summit on Sustainable Development and specifically on the Johannesburg Plan of Implementation (JPI). At the meeting, the JPI established a three-stage hierarchy to guide nations in managing their waste:

1. Prevention and minimization
2. Reuse and recycling
3. Environmentally sound disposal facilities, including technology to recapture the energy contained in waste

The JPI also agreed on the objective of enhancing environmentally sound management of chemicals and hazardous wastes, implementing multilateral environmental agreements, raising awareness of issues, and encouraging the collection and use of additional scientific data in line with the Basel Convention. The plan also seeks to promote efforts to prevent illegal trafficking of hazardous chemicals and hazardous wastes and to prevent damage resulting from the transboundary movement and disposal of hazardous waste (ISWA 2005).

Although the ISWA supported the overall objectives of the JPI, its leaders noted that the waste problem was being viewed from a narrow perspective in relation to sustainable development. "It is important to not only focus on prevention of waste from the end-of-life products, but also try to prevent the so-called hidden flows left behind in other countries." The organization noted that European countries "are increasingly exporting environmental problems, and this trend should be reverted" (ISWA 2005).

Regional Agreements on Hazardous Waste

Numerous regional agreements have been adopted to stem the flow of hazardous waste in addition to the Basel Amendment. In 1991, the 51 members of the Organization for African Unity, now known as the African Union, signed the Ban of Import into Africa and the Control of Transboundary Movement and Management of Hazardous Wastes within Africa, also known as the Bamako Convention, after the city in Mali where it was developed. The agreement bans all waste generated outside of Africa and came into force in 1998. It was enacted because of perceived weaknesses in the Basel Amendment, which many nations believed was not stringent enough. Implementation occurs only if the parties to the convention develop laws prohibiting the importation of hazardous waste from outside Africa through domestic legislation. The Bamako Convention does exceed the provisions of the Basel Amendment in the area of ocean dumping of waste, and it broadens the definition of waste by including radioactive waste and those substances that have been banned, canceled, refused registration by government regulatory action, or voluntarily withdrawn from registration in the country of manufacture for human health and environmental reasons. Enforcement comes from internal regulations and staffing by each party to the agreement, with violators and their accomplices subject to criminal penalties. One of the loopholes in the convention is that it does not ban import into Africa the hazardous waste created by another African nations. It does, however, call for developed nations to transfer hazardous waste processing technology to developing nations, as does the Basel Convention.

The 1990 Lome Convention, in contrast, is a trade agreement. It prohibits the export of hazardous waste from the European Community (EC) to the African, Caribbean, and Pacific (ACP) states. It bans both the direct and indirect export of any hazardous or radioactive waste from European nations to the developing countries within the ACP, who in turn agree not to accept wastes from states outside the EC.

Although it has not ratified the Basel Convention, the United States is a participant in several other international waste activities. In 1986, the United States and Canada signed an agreement to provide both countries with safe, low-cost options for managing waste for which there is either a lack of domestic capacity or the technology to appropriately manage the waste. Another 1986

U.S. agreement with Mexico deals with hazardous waste shipments between the two countries. More limited bilateral agreements have been developed between the United States and Malaysia (1995), Costa Rica (1997), and the Republic of the Philippines (2001) that provide only for the export of hazardous wastes from these countries to the United States (U.S. EPA 2007).

Radioactive Waste

In the 1940s, several nations began using the ocean as a dumping ground for radioactive waste with little concern about the durability of containment vessels. Russia has admitted dumping large amounts of high-level radioactive waste, including nuclear reactors, into the sea since the 1950s. Concerns have also been raised about the aging fleet of Soviet-era nuclear submarines, which have been stored or dismantled with little attention to corrosion and damage to the marine environment. But it took decades after World War II for international regimes on radioactive waste to be approved and implemented.

The Oslo Convention was the first regional treaty to control dumping at sea, negotiated in 1972 by countries bordering the Northeast Atlantic. However, participating countries successfully blocked efforts to include radioactive waste in the agreement. A second treaty, the London Convention on the Prevention of Pollution by Dumping of Wastes and Other Matter, concluded several months later, was the first global dumping agreement, and it did prohibit the dumping of high-level radioactive waste. Dumping is defined as the deliberate disposal of wastes and other matter at sea by ships, aircraft, and man-made structures. The convention covers sewage sludge, dredged materials, construction and demolition debris, explosives, and other materials loaded on a vessel for dumping. Although it is considered one of the most successful treaties covering marine pollution, it has been criticized because it uses the classification of radioactive waste (high, medium, or low level) developed by the International Atomic Energy Commission (IAEA). That system applies to the handling of waste by workers, not its disposal, and concerns have been raised that the classification system might not be applicable at sea.

A 1996 Protocol to the London Convention extensively changed the treaty, banning the incineration of wastes at sea. It also included a provision that intended to prohibit all dumping unless a particular substance was listed in an annex to the agreement and unless a permit was obtained. Parties must show that they attempted to reduce waste at the source, considered other waste management options, and examined environmental effects of dumping. Also required is a duty to manage all waste locally through the prohibition of waste exports to other countries for dumping or incineration at sea.

The dumping of low-level radioactive waste was allowed under the original 1972 agreement if the party obtained an appropriate permit. In 1983, the parties to the convention adopted a nonbinding resolution establishing a moratorium on all marine dumping of radioactive materials while scientific studies were undertaken. Initially, several nations opted out of that portion of the convention, including the United States, the Russian Federation, China, Belgium, France, and the United Kingdom. When the ban was formally implemented in 1994, only the Russian Federation opted out (Hunter, Salzman, and Zaelke 2002, 735).

The 1989 Basel Convention on the Transboundary Movements of Hazardous Waste, while important, did not include radioactive waste. At the time, the IAEA, the UN agency responsible for dealing with radioactive products, did not have any regulations in place for handling waste. It began drafting new guidelines, which were adopted in 1990 by the General Conference of the IAEA, known as the Code of Practice on the International Transboundary Movement of Radioactive Waste. This agreement filled the gap in the Basel Convention that had exempted radioactive waste shipments, and although it is advisory, it does guide parties in taking appropriate steps necessary to ensure radioactive waste is safely managed and disposed of to protect human health and the environment. The code also recognizes the rights of states to prohibit the movement of radioactive waste into, from, or through their territory.

Continuing Waste Management Issues

Electronic Waste

One recent issue involving the waste trade is electronic waste, or e-waste, that is sent abroad. The Basel Action Network and other organizations have addressed the problems that develop when recycling companies in industrialized nations send materials abroad, where it is estimated that 80 percent ends up in poverty-stricken small towns in Asia and Africa that are ill equipped to deal with the waste. Guiyu, China, is considered the center of e-waste recycling; dumping and burning has rendered the local water not potable for drinking, and the river there has 200 times the acceptable levels of acid and 2,400 times the acceptable level of lead. One out of 10 students in the local school has severe respiratory problems.

Although China has banned the import of e-waste, the residents of Guiyu depend on discarded toner cartridges, computers, and printers to earn an average of $1.50 per day. Illicit trade and a lack of enforcement allow recyclers to ship e-waste abroad, where components are taken apart, scavenged, and destroyed. Items like toner cartridges, made of industrial-grade plastic, are almost impossible to melt down and are not biodegradable. Old computers are piled up on top of one another and used as fences, while toxic smoke fills the air as keyboards and cartridges are burned. Although consumers believe that the computer monitor they take back to a store will be recycled in an environmentally safe way, they have no way of knowing that an exporting recycler can make money by sending the equipment to another country. EPA rules and international agreements fail to regulate this type of trade, which means that although e-waste may be diverted from landfills, it still can end up in Africa, Eastern Europe, or Latin America (Judge 2004).

In Africa, much of the e-waste is processed through Lagos, Nigeria, the continent's largest port. The Basel Action Network estimates that 500 shipping containers loaded with secondhand electronic equipment pass through Lagos each month, each packed with a load equal in volume to 800 computer monitors or processors, or 350 large television sets. Although estimates of how much of this equipment is useless waste vary from 25 to 75 percent, it can be assumed that the volume of products entering Lagos equals at least 100,000 computers or 44,000 television sets each month. Other e-waste comes through the port cities of Mombasa, Kenya; Dar es Salaam, Tanzania; and Cairo (Schmidt 2006).

The waste enters Africa for purely economic reasons. Scrap components can be reused by reassembling them, and the cost of shipping a container is less than what the scrap will bring on the African market. Rather than importing computers directly, which requires the payment of a trade tariff, scrap classifications are usually measured in pounds, which are much less expensive. An exporter may sell a container that is full of junked electronics with a few high-value items thrown in, requiring importers to take everything. The leftover materials that cannot be recycled include toxic metals and plastic casings that, when burned, emit carcinogenic chemicals. Some of the equipment is used to fill in swamps in Lagos, resulting in ecological damage. Other equipment, even when refurbished, may have a short life before it, too, becomes obsolete or ceases working. At that point, most recipients of e-waste do not have the capacity to properly dispose of used monitors, cell phones, and television sets (Limo 2007).

Shipbreaking

Shipbreaking is the process of dismantling ships at the end of their useful life, usually about 25 to 30 years. In the last 20 years, this practice has become a growing concern because of media coverage, the activities of groups such as Greenpeace, and the market for scrap metals. This continuing issue is one that has become more acute as the number of vessels being taken apart increases and because of the role of developing countries, especially India, Bangladesh, Pakistan, and Turkey.

The practice usually involves dismantling ships to recover the steel used in building them. But recycling metal is secondary to concerns about the disposal of toxic contaminants such as asbestos, mercury, lead, and other compounds that make up about 5 percent of the ship's weight. Prior to the 1970s and the advent of hazardous waste as a major environmental issue, most shipbreaking occurred in the United Kingdom, Taiwan, Spain, Mexico, and Brazil. Dismantling was primarily a mechanized process. But in recent years, shipbreaking has moved to poor Asian countries where more worker exposure to toxins takes place but fewer concerns are voiced about safety and waste disposal (Clapp 2001).

Greenpeace estimates that out of the 62,000 ships in the world fleet in 2000, about 600 to 700 larger sea vessels are taken out of service each year and brought to Asia for scrap. Some of the ships are single-hull oil tankers that are being phased out by the International Maritime Organization after an oil tanker collapsed in 1999 and leaked heavy oil along the French coast. Another incident took place in 2002 that polluted the coast of Spain and France, causing the European Union to accelerate the phaseout of single-hull tankers. More tonnage is being scrapped each year, mostly by the more than 100,000 workers worldwide who earn between $1.50 and $2.00 per day cutting the metal with torches and carrying heavy loads on their back (Greenpeace 2007).

Uncontrolled Landfills and Scavenging

The environmental and public health issues associated with open dumping are serious problems in developing countries. They pose risks ranging from the contamination of groundwater and soil to injuries related to scavenging. Critics note the detrimental health effects of scavenging, the reduction of efficiency in overall waste management, and the involvement of children in a dangerous activity.

Garbage scavenging can be viewed from several perspectives. In developing countries, a lack of capital for high-tech waste management solutions makes scavenging or trash picking an accepted practice that reduces the amount of waste that needs to be processed and provides an income source for the poor. Some local industries are dependent upon the availability of secondary raw materials for processing, especially metals and fiberboard. In urban areas, scavenging often occurs before waste collection trucks arrive, as residents seek out items like aluminum cans or metal. More commonly, scavenging is done at the disposal site, which is likely to be an open dumping area where some people also live. Martin Medina of the Colegio de la Frontera Norte in Mexico contends that scavenging has existed for centuries and is not a new phenomenon resulting from expanding urban centers. What has changed is the magnitude and visibility of scavenging, especially in cities that do not have the means to collect, process, and dispose of garbage (Constance 2005).

Garbage picking is considered hazardous because individuals seldom work with companies or contracts and do not have health coverage or safety equipment. Criminal elements, the "garbage Mafia," control garbage and recycling in some cities, and many use intimidation or violence to control their territories and services to residents. Some are illegal immigrants, others are not educated, and little police protection is available for those who live and work in the dumps.

Despite these problems, the majority of governments have looked the other way and have allowed or tacitly supported scavenging. In countries such as Egypt, where ragpicking is considered an important part of the informal economy, outlawing the practice would increase poverty levels among those who are already poor. For Malaysia and Thailand, scavenging is an accepted part of the country's culture and one way to assist communities in dealing with the increasing mountains of waste that go hand in hand with emerging development.

References

African Development Bank. 2002. *Study on Solid Waste Management Options for Africa.* July. [Online report; retrieved 8/12/07.] www.afdb.org/pls/portal/url/ITEM/F5F4CC9E2105E31EE030A8C0668C631A.

Basheer, K. P. M. 2007. "Hostility to Garbage Plant Brewing." *The Hindu,* July 12. [Online article; retrieved 8/3/07.] http://www.hindu.com/2007/07/12/stories/2007071259110300.htm.

Cairney, T., and D. M. Hobson, eds. 1998. *Contaminated Land: Problems and Solutions,* 2d ed. London: E & FN Spon.

Clapp. Jennifer. 2001. *Toxic Exports: The Transfer of Hazardous Waste from Rich to Poor Countries.* Ithaca, NY: Cornell University Press.

Colombo, Francesca. 2003. "Mafia Dominates Garbage Industry." *Tierramerica,* June 23. [Online article; retrieved 6/10/07.] http://www.tierramerica.info/nota.php?lang=eng&idnews=1613&olt=224.

Constance, Paul. 2005. "The Transfiguration of Trash Pickers in Brazil and Latin America." *Brazzil Magazine,* May 30. [Online article; retrieved 5/29/07.] http://www.brazzil.com/2005-mainmenu-79/129-may-2005/9293.

Earthtimes. 2007. "Australia Finds Dump Site for Nuclear Waste." *Earthtimes,* May 25. [Online article; retrieved 6/1/07.] http://www.earthtimes.org/articles/show/66232.html.

Ekelund, Lotten, and Kristina Nystrom. 2007. *Composting of Municipal Waste in South Africa.* Uppsala, Sweden: Uppsala University Technical Study.

Environmental News Service. 2007. "Japan Accused of Breaching Toxic Waste Trade Treaty." March 14. [Online article; retrieved 5/22/07.] http://www.ban.org/ban_news/2007/070314_accused_of_breaching.html.

Europa. 2007. "Waste Management." [Online information; retrieved 5/22/07.] http://europa.eu/scadplus/leg/en/s15002.htm.

European Commission. 2007. "Waste." [Online information; retrieved 5/22/07.] http://ec.europa.eu/environment/waste.

Feshbach, Murray. 1995. *Ecological Disaster: Cleaning Up the Hidden Legacy of the Soviet Regime.* New York: Twentieth Century Fund.

GrassRoots Recycling Network. 2000. "Waste Not Asia." August. [Online article; retrieved 5/22/07.] http://www.grrn.org/zerowaste/articles/waste_not_asia.html.

Greenpeace. 2007. "Shipbreaking." [Online information; retrieved 6/7/07.] www.greenpeaceweb.org/shipbreak.

Hawley, Charles. 2007. "Europe's Nuclear Waste Conundrum." *Spiegel Online,* April 19. [Online article; retrieved 8/11/07.] http://www.spiegel.de/international/germany/0,1518,,478309,00.html.

Hiney, Steve, and Ted Hawley. 2005. "Saipan's Solid Waste Management System." *Government Engineering* (November/December): 13–15.

Hunter, David, James Salzman, and Durwood Zaelke. 2002. *International Environmental Law and Policy,* 2nd ed. New York: Foundation Press.

International Solid Waste Association (ISWA). 2005. *ISWA 10 Years Perspective on Waste Management.* Copenhagen, Denmark: ISWA.

Israel Ministry of the Environment. 2007. "Solid Waste." [Online article; retrieved 8/12/07.] http://sviva.gov.il/bin/en.jsp?enPage=e_BlankPage&enDisplay=view&enDispWhat=Zone&enDispWho=waste&enZone=waste.

Judge, Tricia. 2004. "Guiyu Revisited: Exposing the Fraud . . . 18 Months Later." *Imaging Spectrum Magazine* (January): 1–8.

Kloosterman, Karin. 2007. "Israeli Discovery Converts Waste into Clean Energy." *Israel21c.* [Online article; retrieved 8/12/07.] http://www.israel21c.org/bin/en.jsp?enZone=Technology&enDisplay=view&enPage=BlankPage&enDispWhat=object&enDispWho=Articles%5El1586.

Krishna, Gopal. 2006. "Burning Biomass Is Not Green." *India Together,* October 4. [Online article; retrieved 8/7/07.] http://www.indiatogether.org/2006/jul/env-timarpur.htm.

Limo, Andrew. 2007. "Dangers of Increasing Electronic Waste Dumping by Rich Nations." *The Nation* (Nairobi), May 19. [Online article; retrieved 5/29/07.] www.allafrica.com.

Luo, Michael. 2006. "The Danger in the Trash of Baghdad." *International Herald Tribune,* October 11. [Online article; retrieved 5/29/07.] http://www.iht.com/bin/print_ipub.php?file=/articles/2006/10/11/newstrash/php.

Muhumed, Malkhadir M. 2007. "Kenya Dump Poisons Kids, U.N. Study Says." *Arizona Republic* (October 6): A23.

Murdoch, Lindsay, and Jasmin Afianos. 2007. "Nuclear Dump for NT under $12M Deal with Aboriginals." *Brisbane Times,* May 26. [Online article; retrieved 6/1/07.] http://www.brisbanetimes.com.au/news/national/nuclear-dump-for-nt-under-12m-deal-with-aboriginals/2007/05/25/1179601710887.html.

Nadeau, Barbie. 2007. "Italy: Naples's Stinky, Dirty Trash Crisis." *Newsweek,* May 23.

O'Neill, Kate. 2000. *Waste Trading among Rich Nations: Building a New Theory of Environmental Regulation.* Cambridge, MA: MIT Press.

Onishi, Norimitsu. 2005. "How Do Japanese Dump Trash? Let Us Count the Myriad Ways." *New York Times,* May 12. [Online article; retrieved 4/11/07.] http://www.nytimes.com/2005/05/12/international/asia/12garbage.html?scp=1&sq=&st=cse.

Page, G. William. 1997. *Contaminated Sites and Environmental Cleanup: International Approaches to Prevention, Remediation, and Reuse.* San Diego: Academic Press.

Republic of Ireland. 1999. "Consultancy Study on Plastic Bags Recommends Tax on Plastic Bags." [Online news release; retrieved 4/28/07.] www.mindfully.org/Plastic/Laws/Plastic-Bag-Levy-Ireland.

Republic of Ireland, Department of the Environment, Heritage & Local Government. 2007. "The Plastic Bag Levy Will Increase to 22 Cent on Sunday 1 July 2007." February 21. [Online news release; retrieved 4/29/07.] http://www.environ.ie/en/Environment/Waste/PlasticBags/.

Schmidt, Charles. 2006. "Unfair Trade: E-Waste in Africa." *Environmental Health Perspectives* (April): A232–A235.

Suchitra, M., and P. N. Venugopal. 2007. "The Environmental Refugees of Brahmapuram." *India Together,* July 24. [Online article; retrieved 8/7/07.] http://www.indiatogether.org/2007/jul/env-bpuram.htm.

Sule, Surekha. 2004. "Municipalities Overruling the SC." *India Together,* July. [Online article; retrieved 8/7/07.] http://www.indiatogether.org/2004/jul/env-muniswm.htm.

Sule, Surekha. 2005. "Whose Garbage Is It, Anyway?" *India Together,* January 7. [Online article; retrieved 8/7/07.] http://www.indiatogether.org/2005/jan/env-ragpick.htm.

U.S. Agency for International Development (USAID). 2003. *Environmental Issues and Best Practices for Solid Waste Management.* [Online report; retrieved 8/13/07.] http://www.usaid.gov/locations/latin_america_caribbean/environment/docs/epiq/chap5/lac-guidelines-5-solid-waste-mgmt.pdf.

U.S. Environmental Protection Agency (EPA). 2007. "International Waste Agreements." [Online information; retrieved 6/26/07.] http://www.epa.gov/epaoswer/osw/internat/agree.htm.

Zhenhua, Xie. 2007. "Being in Earnest." UN Environment Programme. [Online article; retrieved 6/26/07.] http://www.unep.org/OurPlanet/imgversn/104/zhen.html.

4

Chronology

1757 Ben Franklin institutes the first municipal street-cleaning service to deal with trash in Philadelphia.

1833 Farmers are told that manure is an excellent soil amendment to help make crops grow.

1835 The first bottled soda water is sold in the United States.

1842 A report by the English Poor Law Commission in England concludes that disease is caused by unsanitary environmental conditions, setting the stage for government officials to respond to the problems of urban waste.

1849 New York City police begin a crackdown on residents who own pigs and feed them household garbage despite strict antiswine laws. The "hog problem"—freely wandering swine that eat trash—becomes a symbol of social decay in Manhattan.

1852 Francis Wolle, a botanist, invents and patents a paper bag–making machine, founding the Union Paper Bag Machine Company in 1869.

1863 Dr. Ezra Pulling, a volunteer sanitation inspector in New York City, describes the heinous recycling of city wastes. Rotting food scraps and dead animals are being made into sausage and served at sailors' boardinghouses.

1874 A "destructor" built in Nottingham, England, becomes the first technological garbage incinerator in the world.

1880 An estimated 24 percent of American cities offer some kind of refuse and disposal system.

1885 The first U.S. trash incinerator, called a cremator, is built on Governors Island, New York, by U.S. Army lieutenant H. J. Reilly.

1886 The country's first reduction plant opens in Buffalo, New York, to cook and compress organic wastes using the Merz process, turning them into grease and fertilizer.

1887 The American Public Health Association appoints its Committee on Garbage Disposal.

1888 The U.S. Congress enacts the first federal ban on ocean dumping by passing the Marine Protection Act.

1894 The Massachusetts Institute of Technology offers the first curriculum in sanitary engineering.

1895 George E. Waring is named commissioner of New York's Street-Cleaning Department and is placed in charge of all city street sweeping and trash collection.

1899 The U.S. Patent Office issues the first patent for a glass-blowing machine to produce glass bottles to Michael Owens of the Libby Glass Company.

1906 Dairy owners introduce a one-way paper milk carton to replace glass bottles.

1913 Scrap dealers form the National Association of Waste Materials Dealers (now known as the Institute of Scrap Recycling Industries).

1923 Scarsdale, New York, opens a municipal composting facility where anaerobic bacteria break down organic wastes, transforming them into humus, which is then bagged and sold as fertilizer.

Six-pack soft drink cartons called "Hom-Paks" are created by Ernest Woodruff of the Coca-Cola Company, adding to the packaging of products for home consumption.

1925 A study by the American Child Health Association shows that 10 percent of the cities surveyed still charge their health departments with waste management as municipal governments shift toward street cleansing divisions.

1929 Neighborhood protesters successfully halt the operation of a waste incinerator in San Francisco through a court ruling that deems the facility a public nuisance.

1930 Civil engineers report that all 44 cities in Los Angeles County dispose of their organic waste by selling it to piggeries.

1931 The first International Conference on Public Cleansing is held in London, urging local governments to consider waste management as a recognized technical service.

1932 Milk producers begin providing their product in a one-way, plastic-coated paper carton for consumer use.

1933 The U.S. Supreme Court rules against New York City, deciding that the ocean dumping of municipal wastes is illegal because it creates a public nuisance.

1934 The National Recovery Administration requires a two-cent deposit for small bottles and a five-cent deposit for large ones after bottlers use the noncollection of deposits as a competitive weapon.

The United States' first sanitary landfill is built on the outskirts of Fresno, California, which develops a system whereby garbage is compacted and then covered by dirt to reduce disease transmission and odors.

1935 A brewer in Newark, New Jersey, develops the first aluminum can as a rival to the refillable bottle, but it is not used extensively until the 1950s because of the Great

1935 (*cont.*) Depression, which emphasizes reusing and saving resources, not throwing them away.

1937 John W. Hammes invents the In-Sink-Erator, the United States' first kitchen garbage grinder. Some municipal governments, including New York City, ban the devices for fear they will overtax aging sewage systems.

1939 The U.S. Public Health Services reports that 52 percent of U.S. cities are feeding municipal garbage to pigs.

1941 The American Public Works Association, formed in 1937 with the merger of the American Society of Municipal Engineers and the International Association of Public Works Officials, publishes the first manual devoted to refuse collection. It represents the first serious effort in the United States to consider both the basic collection requirements and the economic considerations in establishing a foundation of best practices in the field.

1943 The U.S. Public Health Service recommends the use of sanitary landfills to municipal governments as a wartime labor and resource conservation measure.

1948 New York City opens Fresh Kills landfill on Staten Island to dispose of the city's household refuse in an efficient, sanitary, and unobjectionable manner pending the building of incinerators. The facility is considered a model for other landfills in the United States.

The U.S. Public Health Service Division of Sanitation includes among its programs departments on air quality, water supply, sewage/wastewater, and radiation, which will eventually be moved to the newly established Environmental Protection Agency in 1970.

1950 Henry Wasylyk, a Canadian inventor from Winnipeg, Manitoba, invents the green plastic garbage bag made from polyethylene, initially intended for hospital use rather than residential use.

1951 Marion Donovan, a mother, receives a patent for the disposable diaper, one of the least recycled items found in municipal solid waste.

1953 Vermont's legislature bans the sale of all throwaway beer bottles in the state. The law is rescinded in 1957, when the state legislature fails to renew the bottle ban, apparently because it did not reduce litter.

1955 *Life* magazine heralds the advent of the "throwaway society."

1957 The National Academy of Sciences, in a study commissioned by the U.S. Atomic Energy Commission (AEC), recommends that highly radioactive waste be stored in deep geologic formations.

Ermal Cleon Fraze invents the pop-top aluminum can and sells the concept to Alcoa Aluminum four years later.

1961 Procter and Gamble introduces the first paper disposable diaper to consumers.

The American Public Works Association's Committee on Refuse Disposal publishes the manual of practice, *Municipal Refuse Disposal.* It includes the term *solid waste* and sections on sanitary landfills, incineration, grinding food wastes, composting, swine feeding, salvage and reclamation, and management practices.

1962 The Pittsburgh Brewing Company markets a pull-ring top for consumer beverages in cans.

1964 One-way plastic jugs made of high-density polyethylene are introduced for milk containers.

1965 The Solid Waste Management Act is passed by the U.S. Congress to stimulate research on waste disposal and reduction of the waste stream by providing technical and financial support for solid waste planning.

1965 (*cont.*) U.S. president Lyndon B. Johnson creates the Commission on Natural Beauty as part of the effort to control litter and promote community beautification.

1967 The U.S. Air Quality Act requires older incinerators to add air pollution controls to reduce soot and fine particles.

1968 Dean Buntrock starts the United States' largest waste hauler, Waste Management, Inc., in Chicago, joining the roster of the New York Stock Exchange in 1971 as one of the first garbage companies to go public.

The U.S. Public Health Service publishes guidelines on hospital waste, warning that legal implications and precautions should be taken to ensure the safety of staff, collectors, and the public.

President Johnson's Scientific Advisory Committee recommends that the federal government take a leadership role in solid waste policy.

1969 The U.S. Atomic Energy Commission conducts experiments to store radioactive waste in an abandoned salt mine in Lyons, Kansas.

Browning-Ferris Industries begins hauling waste in Houston after the company's founder grows disgruntled with his local garbage service. It becomes the first refuse company to expand across North America and the second-largest waste handler in the United States. It is purchased in the late 1990s by Allied.

The New York Sanitation Department demonstrates that the curbside pickup of plastic trash bags is cleaner, safer, and quieter than metal can pickup, leading to a shift to plastic trash can liners by consumers.

1970 The U.S. Resource Recovery Act amends the Air Quality Act and shifts the emphasis from waste disposal to recycling and reuse of materials.

The U.S. Atomic Energy Commission establishes a policy whereby the federal government will accept high-level radioactive waste for long-term disposition, but generators of the waste will be responsible for paying for the full costs once the waste is delivered to the government.

The U.S. Department of Transportation creates the Hazardous Materials Information System to monitor safety issues involving hazardous waste. The system is designed to implement parts of the Hazardous Materials Transportation Control Act of 1970.

Gary Anderson designs a logo for the Container Corporation of America, a paper products manufacturer, that becomes the national symbol for recycling.

1972 After a decade of study, the Atomic Energy Commission determines an underground salt mine near Lyons, Kansas, is unacceptable for use as a repository for radioactive waste. Kansas officials raise technical concerns over the suitability of the site. AEC officials announce they will examine a site in southeastern New Mexico as a potential site, with a goal of opening the facility by 1980. The site was proposed by a group of leaders from Carlsbad, New Mexico.

After 18 months of difficult negotiations under the auspices of the United Nations' (UN) Inter-Governmental Working Group on Marine Pollution, the London Convention on the Prevention of Marine Pollution by Dumping of Wastes and Other Matter is enacted, prohibiting the disposal of both low-level and high-level radioactive wastes and other radioactive matter at sea.

The Oslo Convention (Convention for the Prevention of Marine Pollution by Dumping from Ships and Aircraft) is adopted by 13 Northeast Atlantic Ocean countries to regulate the dumping of wastes along the coastline from Portugal to Norway; it does not cover radioactive wastes.

1972 (*cont.*) Oregon passes the first measure imposing a five-cent refundable deposit on all beer and soda bottles and cans. It also bans pull tabs, effectively making all standard drink cans illegal.

1973 Minnesota's state legislature passes the Packaging Review Act, noting that although recycling is one alternative for conserving natural resources, source reduction is also in the public interest. The state's Supreme Court upholds the law after a concerted effort by container makers to argue the law unconstitutionally interferes with interstate commerce.

Nathaniel Wyeth, brother of artist Andrew Wyeth, creates the polyethylene terephthalate plastic bottle after experimenting with a plastic detergent bottle in his home refrigerator.

1974 The Association of State and Territorial Solid Waste Management Officials is incorporated to support state-level environmental agencies and affect national waste management policies.

The Energy Reorganization Act establishes the U.S. Nuclear Regulatory Commission, directing the agency to develop policies for the civilian use of nuclear energy and its waste disposal.

The EPA warns that wastes from medical patients are often hazardous and that difficulties are involved in segregating these wastes from other, nonhazardous, hospital waste, so all such wastes must be considered potentially contaminated and receive special handling and treatment.

The federal government chooses a location 30 miles east of Carlsbad, New Mexico, for exploratory work and field investigations as a possible nuclear waste repository.

California defeats a bottle bill similar to Oregon's after a strong public relations campaign is mounted against it.

The stay-on tab for canned beverages is invented and introduced by the Falls City Brewing Company of

Louisville, Kentucky. The stay-on tab is considered more environmentally sound because the tabs do not end up littering the environment, as is the case with pop-top or pull tabs.

1975 Concerns about human exposure to toxic chemicals are raised by residents at the former site of the Hooker Chemical Company disposal site at Love Canal, New York.

The Hazardous Materials Transportation Act is enacted to deal with increasing concern over accidents involving the transport of hazardous waste in the United States.

The European Commission develops regulations for the transport, treatment, and disposal of hazardous wastes to avoid harm to human health or the environment. Each member state, however, is given the opportunity to define "hazardous waste" and to develop its own set of regulations.

1976 The U.S. Congress enacts the Resource Conservation and Recovery Act (RCRA) to protect the public from the hazards of waste disposal.

California prohibits the siting of any new nuclear reactors until the federal government demonstrates a means of disposal for high-level radioactive waste.

McDonald's restaurants announce a shift away from paper-based packaging.

The American Can Company opens a facility in Milwaukee named Americology, to separate municipal solid waste into recoverable components. The plant closes because of financial problems from a failed plan to sell processed trash to local utilities as a substitute for coal.

1977 The plastic grocery bag is introduced to the supermarket industry as an alternative to paper sacks.

The U.S. Nuclear Regulatory Commission states that the agency will no longer license nuclear reactors if it does

1977 (*cont.*) not have reasonable confidence that the waste can and will in due course be disposed of safely.

1978 The U.S. Public Utility Regulatory Policies Act requires local utilities to purchase electricity generated by waste-to-energy plants. Subsidies are offered to incinerator firms to allow the United States to move toward "energy independence."

The Minnesota legislature requires that all packaging proposals be filed with a state agency for evaluation as part of state solid waste reduction goals.

The National Recycling Coalition is founded to eliminate waste and promote recycling programs as the U.S. environmental movement highlights promotion of sustainable economies.

U.S. president Jimmy Carter establishes the Interagency Review Group on Nuclear Waste Management to develop federal policy.

The U.S. Supreme Court rules in *Philadelphia v. New Jersey* that garbage is to be treated like any other commercial commodity under the Commerce Clause of the U.S. Consitution, which gives Congress the authority to regulate interstate commerce.

The first test hole is dug at Yucca Mountain, Nevada, as part of a nationwide search for a nuclear waste site.

1979 The U.S. Congress authorizes the Waste Isolation Pilot Plant (WIPP) as a research and development facility to demonstrate the safe disposal of radioactive wastes resulting from defense activities.

1980 Seattle develops a program to encourage composting among city residents as one way to reduce waste.

1981 New Mexico's attorney general files suit in U.S. district court alleging violations of federal and state laws in connection with the WIPP facility.

1982 The U.S. Environmental Protection Agency (EPA) publishes guidelines for infectious waste management.

The U.S. Nuclear Waste Policy Act seeks long-term permanent storage, rather than retrievable storage, of nuclear waste, and Congress directs the Department of Energy to nominate at least five sites for repository selection, with three sites to be recommended to the president.

The U.S. Department of Energy lays out a detailed process and schedule for geologic repositories for spent nuclear fuel and high-level radioactive waste.

1983 A study by the National Academy of Sciences recommends on-site source reduction by chemical companies and treatments to reduce the toxicity of hazardous chemical waste.

In response to passage of a bottle bill exempting aluminum cans, the beverage industry establishes the $20,000 Bovine Trust Fund to reimburse a farmer for any cow or horse certified to have died from eating an aluminum can.

The U.S. Department of Energy identifies nine potentially acceptable sites for long-term storage of radioactive waste; six are located in the West and three in the South.

The U.S. Nuclear Regulatory Commission promulgates high-level radioactive waste regulations.

1984 The member states of the Organisation for Economic Cooperation and Development (OECD) agree to the Decision and Recommendation on Transfrontier Movements of Hazardous Waste, the first international agreement regarding the international trade of hazardous waste to ensure authorities in countries affected by shipments are provided adequate and timely information on waste movements.

In amending and reauthorizing RCRA, the U.S. Congress declares its intention to promote recycling, waste

1984 (*cont.*) reduction, and environmentally sound waste management practices.

1985 The Ocean Conservancy conducts a study for the Environmental Protection Agency and finds plastics are the number one marine debris hazard.

The EPA sets generally applicable environmental standards for high-level radioactive waste disposal.

The California Waste Management Board warns that the state will run out of landfill space by 1997 and recommends increased use of incineration as the solution, with a minimal role for recycling.

1986 The Citizens' Clearinghouse for Hazardous Wastes, founded by Lois Gibbs, starts its Solid Waste Action Project.

U.S. president Ronald Reagan signs the Superfund Amendments and Reauthorization Act (SARA) as a bipartisan measure to set up an $8.5 billion fund to clean up waste sites.

In an attempt to rid the city of its incinerator ash, Philadelphia loads up a cargo ship, the *Khian Sea,* in what would later be called an incident of "waste imperialism." The ash apparently is dumped somewhere at sea.

The GSX waste handling company refuses to pick up waste from eight Boston hospitals because area landfills will no longer accept them. The incident focuses attention on commingled waste issues.

Prior consent guidelines on the transboundary shipment of hazardous waste are extended to OECD members and nonmember states. The guidelines extend to non-OECD states that lack proper hazardous waste disposal facilities.

Responding to a warehouse fire, New York City officers discover 1,400 bags of medical waste that have been illegally dumped there despite documents showing the waste had been incinerated.

The U.S. Department of Energy identifies three sites as suitable for the storage of nuclear waste: Hanford, Washington; Yucca Mountain, Nevada; and Deaf Smith County, Texas.

1987 The UN Environment Programme's Governing Council approves the nonbinding Cairo Guidelines on Environmentally Sound Management of Hazardous Wastes, a precursor to a global convention on waste.

A garbage barge, the *Mobro 4000*, sails from Long Island, New York, in an unsuccessful 164-day effort to find a place to unload more than 3,100 tons of waste cargo from the city of Islip, New York. The trash odyssey takes the barge on a 6,000-mile journey before it returns to New York, where the trash is incinerated and buried where it began.

Greenpeace officially launches a campaign to end the international trade in hazardous waste.

Children in Indianapolis are found playing with vials of blood infected with AIDS that they found in a trash bin outside a health clinic. Officials report the practice of dumping medical waste in bins is legal.

The Nuclear Waste Policy Amendments Act directs the U.S. Department of Energy to study only one site as a repository for nuclear waste: Yucca Mountain, Nevada.

A federal Circuit Court of Appeals decision directs the EPA to modify its regulations for high-level radioactive waste, established in 1985, to ensure they are consistent with existing environmental laws; standards are not revised, however, until 1993.

The mayor and members of the Los Angeles City Council cancel plans to build three trash incinerators after lobbying by the Concerned Citizens of South Central Los

1987 (*cont.*) Angeles, who are concerned about the health effects of toxins released by the plant.

A report by the United Church of Christ Commission on Racial Justice documents that an area's racial composition is the single most reliable factor in determining the location of waste disposal sites.

1988 A garbage slick nearly a mile long is discovered along the shore of Ocean County, New Jersey, as needles, syringes, and empty prescription bottles wash up on local beaches. Six weeks later, 10 miles of Long Island, New York, beaches are closed when medical wastes wash ashore, and similar incidents are reported throughout the summer.

The U.S. Office of Technology Assessment releases a report that finds inconsistencies in federal guidelines for states regarding the definitions and management options for medical and infectious waste. The report notes that no federal regulations comprehensively address the handling, transportation, treatment, and disposal of medical waste.

President Reagan signs the Medical Waste Tracking Act as an important step forward in the protection of the environment and public health.

African nations meet in Monrovia, Liberia, with UN officials and representatives of nongovernmental organizations to recommend a ban on the movement of toxic waste to Africa. The meeting leads to the adoption of Resolution 1153 by the Organization of African Unity to condemn waste imports.

Delays in opening the WIPP in New Mexico prompt Idaho's governor to impose a ban on out-of-state waste shipments to the Idaho National Engineering Laboratory, although the state subsequently agrees to allow shipments to resume temporarily.

The EPA submits a report to Congress on the adequacy of the agency's solid waste guidelines and criteria for protecting human health and the environment, concluding that current federal, state, and local regulations are not addressing the problem.

1989 The first International Coastal Cleanup is held to collect trash from coastal states in the United States, Canada, and Mexico.

A U.S. firm ships 200 barrels of hazardous waste to Zimbabwe using false labels that describe it as "dry cleaning fluids and solvents," in a well-publicized case that shows exports of hazardous waste are becoming routine.

U.S. Department of Energy secretary James Watkins announces an indefinite delay in opening the WIPP, saying the project will open when DOE's safety concerns are addressed and other key reviewers are satisfied.

1990 The McDonald's restaurant chain abandons polystyrene (Styrofoam) containers in favor of quilt-wrap to reduce criticism of its environmental practices. The fast-food restaurant chain had been the object of nationwide protests, including a "send it back" campaign of used clamshell-style food containers. The company had described Styrofoam as "basically air" that was good for landfills because it aerated the soil.

African, Caribbean, and Pacific states, including former colonies of Western European powers, sign the Lome Convention, prohibiting the export of hazardous or radioactive waste from the European Community to the countries within these regions. The parties agree not to accept waste imports from any other states outside the European Community, effectively halting developed country waste shipments, the parties believe.

The General Conference of the International Atomic Energy Agency adopts the Code of Practice on the International Transboundary Movement of Radioactive Waste to fill the gap that exempted radioactive waste shipments from coverage under the Basel Convention.

1991 Germany enacts a landmark packaging ordinance that shifts the burden of collecting, sorting, recycling, and disposing of waste away from consumers and onto manufacturers in a version of "extended producer responsibility."

Twelve African countries meeting in Bamako, Mali, sign the Bamako Convention to ban the importation of hazardous waste from any country; the convention comes into force in 1998.

The EPA adopts new regulations for municipal incinerators, requiring operators to use New Source Performance Standards and issuing guidelines for existing facilities.

U.S. regulations developed under the Medical Waste Tracking Act of 1988 expire after pilot programs establish a "cradle to grave" tracking system.

The Protocol on Environmental Protection to the Antarctic Treaty (Madrid Protocol) is adopted to regulate the incineration and removal of waste materials from the continent.

The U.S. Department of Transportation publishes a final rule to define and control regulated medical waste and medical waste to distinguish waste consisting of an infectious substance.

The first National People of Color Environmental Leadership Summit is held, with toxic waste dumping and waste disposal facility siting as its key issues.

Secretary of Energy Watkins notifies the Department of Interior that the WIPP is ready to begin accepting radioactive wastes for a five-year test phase. New Mexico, Texas, and environmental organizations file suit asking that the DOE first receive congressional approval before wastes are sent to the WIPP, and a judge agrees, granting an injunction.

The EPA announces new regulations governing landfills, requiring operators to monitor nearby groundwater for the presence of 70 pollutants, to use double liners of flexible materials, and to have a leachate collection system.

1992 The Basel Convention on the Control of Transboundary Movements of Hazardous Wastes and Their Disposal, convenes under the auspices of the UN Environment Programme, enters into force to deal with the issue of waste trading among nations.

The Convention for the Protection of the Marine Environment in the North East Atlantic replaces the Oslo and Paris conventions and requires parties to show that no harm to the environment will be caused by dumping waste into the North Sea and that no adequate disposal alternative exists.

Six Central American countries sign the Agreement on the Transboundary Movement of Hazardous Wastes in the Central American Region following a strong campaign by environmental organizations.

The National Energy Policy Act requires the U.S. Department of Energy to continue to oversee the Yucca Mountain nuclear waste site to prevent activities that could adversely affect the repository. The statute also directs the EPA to prepare a high-level waste standard specifically for Yucca Mountain and provides for the National Academy of Sciences to make recommendations to the EPA on the scientific basis for the health and safety standards for Yucca Mountain.

In *Fort Gratiot Landfill v. Michigan Department of Natural Resources,* the U.S. Supreme Court rules that the goals of solid waste management plans can be attained without discriminating against interstate commerce.

U.S. president George H. W. Bush signs legislation establishing conditions for initial receipt and permanent disposal of waste at WIPP, designating the EPA as the site's independent regulator.

1993 Nineteen countries in the Mediterranean region negotiate an agreement to ban hazardous waste trading under the framework of the 1975 Barcelona Convention for the Protection of the Mediterranean Sea against Pollution.

The first criminal prosecution under the 1980 RCRA involves the conviction of two men for knowingly exporting hazardous waste from the United States to Pakistan without the required consent and without notifying the EPA prior to shipment.

The Environmental Defense Fund criticizes plastic manufacturers, charging that the rate of recycling does not even come close to the increased production of virgin plastic.

The EPA reports that domestic recycling has tripled by weight, from 7 percent in 1970 to 22 percent, with more cities passing voluntary and mandatory recycling measures.

Browning-Ferris Industries releases first contract bids to haul trash in New York City. Shortly thereafter, a company executive opens the door to his home to find the severed head of a German shepherd with a note in its mouth that reads, "Welcome to New York," allegedly left there by the Mafia as a warning.

The Oregon Supreme Court rules that a statute authorizing higher fees for disposal of out-of-state waste does not on its face violate the Commerce Clause; the U.S. Supreme Court overturns the Oregon decision a year later in a 7-2 decision calling the surcharge discriminatory on its face.

1994 The North American Free Trade Agreement goes into effect and includes provisions related to hazardous waste, including agreement that relaxed environmental standards will not be used to attract foreign investment.

1995 The Waigani Convention is adopted in Waigani, Papua New Guinea, to address the dumping of waste in the

South Pacific region. It bans the import of hazardous waste and radioactive waste from outside the convention area to developing countries in the convention area.

The Smoky Mountain Garbage Dump, home to an estimated 150,000 people and the source of fires and landslides, is closed in Manila, Philippines, after international pressure makes the facility an embarrassment to the government.

The U.S. Congress enacts legislation to provide a storage facility for spent nuclear fuel near the Nevada Test Site, because an expected repository at Yucca Mountain is not yet available.

The Ocean Conservancy produces a report on ship waste and discharge at sea to call attention to the problems of oceanic trash.

The U.S. Department of Transportation publishes a final rule dealing with requirements for packaging and labeling regulated medical waste in response to waste management industry concerns about regulatory burdens.

The National Research Council publishes recommendations on health and safety standards for the Yucca Mountain nuclear waste repository.

The New York Police Department completes an extensive undercover operation that officially wipes out the city's garbage cartel with a 114-count indictment, and every individual named and charged either pleads or is found guilty.

1996 The Izmir Protocol of the Barcelona Convention is signed by Mediterranean coastal states to regulate waste trade in the Mediterranean region. The agreement is based on a draft proposal developed by Greenpeace.

The Protocol to the Convention on the Prevention of Marine Pollution by Dumping of Waste and Other Matter is adopted, revising most of the provisions of the 1972

1996 (*cont.*) London Dumping Convention. It bans the incineration at sea of wastes and requires parties to keep records showing the nature and quantities of permitted dumping as well as the location, time, and method used.

New York City Mayor Rudolf Giuliani establishes the Trade Waste Commission, giving city officials sweeping powers to reject waste handling licenses, whether a firm is criminally charged or not. The action further transforms the city's trash business.

1997 A South African government inquiry finds that the mercury waste facility operated by Thor Chemicals is "out of control" as waste is improperly stored in warehouses at the plant site.

New York City legalizes the use of in-sink garbage grinders for disposing of food waste.

The coordinator of the Greenpeace campaign against hazardous waste trading establishes the Basel Action Network to develop a grassroots network of locally based groups to persuade governments to ratify the Basel Convention.

The Joint Convention on Safety of Spent Fuel Management and on the Safety of Radioactive Waste is adopted in Vienna, relying on the International Atomic Energy Agency for its secretariat.

1998 The EPA announces its certification of the WIPP and compliance with federal standards for disposal.

Unable to get a permit for disposal, a Taiwanese plastic firm ships mercury-contaminated waste to Cambodia. The sacks of waste are mislabeled as polyester chips and cement cake and are looted by local residents, resulting in several deaths. The waste is eventually shipped back to Taiwan.

After a two-year study, the U.S. Department of Energy concludes that Yucca Mountain remains a promising site for a nuclear waste repository. The department fails, however, to meet a deadline for being able to accept spent nuclear fuel; lawsuits are subsequently filed by utilities and states with reactors.

New York repeals its one-cent tax on nonrefillable containers of carbonated soft drinks, mineral water, and soda water.

1999 The Declaration and Decision on Environmentally Sound Management of Hazardous Wastes is adopted in Geneva as an outgrowth of the Basel Convention, setting priorities for the agreement and focusing attention on waste management within member countries.

The Philippine government announces that a Japanese firm has shipped hazardous and medical waste, labeled as paper for recycling, for disposal in Manila in direct violation of the Basel Convention. An embarrassed Japanese government removes the waste; the owner of the Japanese company commits suicide.

U.S. Department of Energy secretary Bill Richardson sends the first shipments of radioactive waste to the WIPP from the Los Alamos National Laboratory.

2000 U.S. presidential candidate George W. Bush tells Nevada governor Kenny Guinn that "sound science, not politics" would guide his decision on Yucca Mountain as a site for a nuclear waste repository.

Twenty-six Greenpeace activists are arrested in Manila after they deliver a container filled with poisonous chemical waste collected in residential areas near Clark Air Force Base to the U.S. Embassy. The protesters called upon the U.S. government to clean up its military waste; the United States withdrew from the Philippines in 1992, leaving its bases in their current state.

Russia's Atomic Energy Ministry wins preliminary approval from the lower house of parliament to import spent nuclear fuel rods. The spent fuel is to be sent by armored

2000 (*cont.*) train to a facility in the Ural Mountains for reprocessing, earning the country as much as $20 billion.

2001 The Fresh Kills landfill on New York's Staten Island receives its final shipment of household garbage on March 22, responding to citizen complaints and government reports about the environmental impact of the landfill's operation and a 1996 state law requiring closure of the facility by January 1, 2002.

Debris from the September 11 terrorist attack on the World Trade Center in New York City is sent to Fresh Kills landfill to be sifted for human remains.

The EPA sets safety standards for the Yucca Mountain nuclear waste facility. An interim agency report finds that no "showstoppers" arose in its scientific review of the site. The estimated cost for construction, operation, and monitoring over 100 years is put at $58 billion.

The U.S. Department of the Interior's National Park Service moves to make the Fresno, California, municipal landfill the first to use a trenching method of covering trash with dirt every day, a national historic landmark. On the same day, the site is placed on the EPA's list of federal Superfund sites for cleanup.

2002 Waste management targets are included as part of the Johannesburg Plan of Implementation at the World Summit on Sustainable Development.

Ireland institutes a 20-cent-per-bag tax on plastic bags, leading to a 95 percent reduction in their use as consumers bring their own bags for shopping.

U.S. secretary of energy Spencer Abraham recommends Yucca Mountain to President George W. Bush as suitable for further development as a nuclear waste repository. President Bush then recommends the site to Congress as qualified for application for construction authorization. The state of Nevada vetoes the recommendation as it is permitted to do under the Nuclear Waste Policy Act, but Congress overrides Nevada's veto and approves the site.

The state of Nevada bans the U.S. Department of Energy from drawing water from wells at the Yucca Mountain nuclear waste facility in an attempt to shut the project down. Federal officials counter that the project will draw instead from a 1-million-gallon water tank. The state implemented a similar shutoff in 2000, but the federal government successfully sued to keep the water flowing.

Greenpeace activists are arrested when they attempt to clean up toxic waste in Bhopal, India.

The U.S. Department of Transportation publishes a final rule that adopts the World Health Organization definitions on hazardous materials for packaging and transport to standardize U.S. regulations with those recommended by the United Nations.

2004 The National Commission on Energy Policy issues a report urging completion of a national underground repository for nuclear waste but also recommends the government move ahead on a parallel path of building at least two centralized dry cask storage facilities at reactor sites.

The U.S. Court of Appeals sets aside an EPA regulation that established a 10,000-year standard for high-level nuclear waste because it differs from a National Academy of Sciences finding that no scientific basis exists for that time limit.

2005 San Francisco enters into voluntary agreement with lobbyists for grocery stores to stop using plastic bags.

The EPA issues draft regulations establishing a two-tier standard for radiation protection at the Yucca Mountain nuclear waste site.

Department of Energy officials admit that a U.S. Geological Survey scientist fabricated documents related to the

2005 (*cont.*) Yucca Mountain nuclear waste site, jeopardizing the operating license application before the Nuclear Regulatory Commission.

The Nevada Commission on Nuclear Projects declares the Yucca Mountain repository a "dead man walking" and predicts the project can be killed.

2006 The U.S. Department of Energy announces it intends to relocate a 130-acre, 12-million-ton pile of radioactive waste in southeastern Utah where it sits 750 feet from the Colorado River, the water source for 25 million people. The waste began piling up in the 1950s, when uranium was mined in the area.

California implements a ban on placing batteries, electronic devices, and fluorescent lightbulbs in household trash because of concerns the wastes will leak toxic substances, mostly metals, into landfills.

2007 San Francisco outlaws plastic checkout bags at large supermarkets and chain pharmacies after a voluntary system proves to be insufficient.

An Israeli firm announces it has developed a new technology for reducing waste, including radioactive waste, and turning it into glass blocks and energy, which can be sold. The company plans to use the technology to dispose of radioactive waste from the Chernobyl nuclear waste accident.

Greenpeace activists dump poisonous waste collected from the site of the 1984 Bhopal, India, gas leak disaster at the European headquarters of Dow Chemical Company to bring attention to the company's failure to clean up the facility.

An Indian firm claims that cleaning up the 350 tons of tarry hazardous waste from the Bhopal gas leak in 1984 is "child's play" despite environmental groups' concerns that the company is not equipped to incinerate the material.

Naples, Italy, faces a garbage crisis as uncollected trash piles up on city streets and washes up on beaches. Plans to build four new dumps in the city are stymied by residents who do not want incinerators in their neighborhoods and local crime syndicates who do not want the new facilities to cut into their illegal dumping profits. Garbage collectors contend they have no place to take the trash.

Lady Bird Johnson, the former first lady, dies. During Lyndon B. Johnson's administration, she became one of the nation's first advocates for highway beautification, introducing stop-littering campaigns across the United States.

Los Angeles and San Francisco ban city-financed purchases of bottled water in an effort to cut down on plastic waste.

2008 Scotland's government issues plans for a zero-waste strategy in order to rid itself of the "dirty man of Europe" label.

Chicago becomes the first city in the United States to tax bottled water, levying a 5-cent-per-bottle tax that is expected to raise nearly $11 million a year in revenue for the city. The International Bottled Water Association files suit to overturn the tax, arguing the city's ordinance unlawfully taxes a food product, a violation of state law.

The government of China bans the production, sale, and use of ultrathin plastic bags after June 1, 2008, and encourages shoppers to use fewer bags or use cloth bags instead. Supermarkets and shops are banned from giving free plastic bags to customers but are allowed to sell them.

Chicago ends its controversial blue bag recycling program amid complaints from citizens concerned about the durability of the bags. City officials announce they will begin distributing blue plastic carts for recycling as an alternative.

5

Biographical Sketches

Robert Bullard (b. 1946)

As "the father of environmental justice," Robert Bullard is both an academician and an activist who is known for his pioneering work in studying the siting of hazardous waste facilities in communities of color and poverty. He is Ware Professor of Sociology and director of the Environmental Justice Resource Center at Clark Atlanta University. Previously, he taught at the University of California, Riverside, and was visiting professor in the Center for African American Studies at the University of California, Los Angeles.

A native of Alabama, Bullard attended Alabama A&M University, receiving a bachelor's degree in government in 1968, then joining the U.S. Marine Corps for a four-year stint as a communications specialist. He went to graduate school at Clark Atlanta University, where he earned a master's degree in sociology in 1972, and in 1976, he earned a PhD in sociology from Iowa State University. His work as a sociologist became focused when he completed graduate school and his wife, an attorney representing a group of homeowners, asked him to help collect data about a company that had decided to build a landfill in a predominantly black neighborhood in Houston. Bullard was an assistant professor in Houston and familiar with the area. His research found that 100 percent of all the city-owned landfills, and six of its eight incinerators, were in black neighborhoods, even though Houston's population was only 25 percent black.

Bullard's best-known book is *Dumping in Dixie: Race, Class and Environmental Quality,* which has become a classic text on environmental racism and justice issues. He served as one of the planners for the First National People of Color Environmental Leadership Summit and served in President Bill Clinton's transition team and on the U.S. Environmental Protection Agency (EPA) National Environmental Justice Advisory Committee. Since his first book appeared, dozens of law centers specializing in environmental justice have been developed, and hundreds of grassroots organizations have formed to oppose the types of hazardous waste sites and landfills Bullard initially described in 1990. He has helped organizations to network on strategies and information to collaborate against discrimination in many forms. In 1994, he founded the Environmental Justice Research Center at Clark Atlanta University, bringing together the expertise that has developed since his initial work on hazardous waste siting began in 1979.

Dean Buntrock (b. 1931)

Dean Buntrock was born in Columbia, South Dakota, a town that had a population of 250 at the time of his birth. At age nine, he started working for his father's local farm equipment dealership, which included a hardware store and gas station. By the time he was 13, he was driving a grain hauling truck and earned money raising chickens and growing potatoes for local restaurants. He attended a one-room Lutheran school and graduated from a high school class of eight. He joined the army in 1952 and attended finance school at Fort Benjamin Harrison. When he was discharged, he attended St. Olaf's College and graduated in 1955, majoring in business and history.

He and his wife moved to Boulder, Colorado, where he sold life insurance. In 1956, when his father-in-law died, he helped manage the family's garbage truck business. In 1968, he joined Wayne Huizenga and Larry Beck in forming Waste Management, Inc., taking the company public in 1971 and leading it to become a major U.S. and global corporation. In the course of a decade, Waste Management had acquired hundreds of garbage hauling companies, making it the largest waste management company in the world. In 1993, Waste Management, Inc., became WMX Technologies; Buntrock retired as chief executive officer in 1997.

Buntrock has a history of public and philanthropic service. He has served on the boards of directors of numerous organizations, including the Chicago Symphony Orchestra, St. Olaf

College, and the First National Bank of Chicago. His honors include an honorary degree from Knox College and membership in the Horatio Alger Association of Distinguished Americans. In a controversial 1987 move that drew protests from activists enraged that an industry insider would be named to the board of directors of the National Wildlife Federation, charges were made that Buntrock had bought his company's way into environmental respectability. Despite its financial successes, Waste Management has had a troubled history that includes accusations and lawsuits alleging price fixing and a federal antitrust case.

Joseph R. Cerrell (b. 1935)

In 1966, Joseph R. Cerrell and his wife, Lee, founded a company that is now one of the United States' oldest government affairs, public relations, and campaign management firms, Cerrell Associates of Los Angeles. In 2007, he was inducted into the American Association of Political Consultants Hall of Fame as one of the most respected and admired figures in the profession. Cerrell Associates has a unique place in the annals of waste management that is rarely mentioned in biographical profiles, but it is important because of a controversial report he authored dealing with the siting of hazardous waste facilities.

Joe Cerrell was born in New York City and attended the University of Southern California; he also studied at Columbia University. Prior to starting his company, he was the youngest-ever director for the California Democratic Party and worked for numerous political candidates from the local level to the presidential level. In his lengthy career, he has served as an officer in organizations ranging from the Public Relations Society of America to the National Italian American Federation. He cofounded the Jesse M. Unruh Institute of Politics and is a Distinguished Visiting Professor at Pepperdine University.

In 1984, the California Waste Management Board commissioned a $33,000 study by Cerrell Associates to identify the political difficulties inherent in siting waste-to-energy plants. The report found that all socioeconomic groupings tend to resent the siting of nearby waste facilities, but that middle- and higher-income neighborhoods are better able to resist these types of projects. The recommendation of the infamous Cerrell Report was what some refer to as class targeting in siting waste facilities. Ideally, the report noted, officials and companies should look for lower socioeconomic neighborhoods in order to minimize opposition to facilities. The study also said that political criteria were as important as engineering factors in determining the outcome of a project, and that political problems could be avoided if supporters and opponents could be identified before a site for an incinerator was selected. The report even identified a personality profile of those likely to offer the least resistance to siting decisions: older people, people with a high school education or less, and those who adhere to a free market orientation. In January 1988, the *Los Angeles Times* broke the story of the previously unpublicized Cerrell Report, triggering a backlash against the waste industry and charges that the 87-page document was the "smoking gun" that proved environmental racism among officials.

Rev. Benjamin Chavis Jr. (b. 1948)

Benjamin Chavis Jr., who was born in Oxford, North Carolina, represents a family tradition of social activism. His parents were teachers at a school for black orphans and taught him about racial injustice at an early age. When he was 13, he tried to integrate the all-white library where he lived and became the first black person to obtain a library card there. After high school, he went to Saint Augustine College in Raleigh, North Carolina, where he became active in the Southern Christian Leadership Conference as a youth coordinator. He received a bachelor's degree in chemistry from the University of North Carolina, Charlotte, in 1969 and was appointed Southern Regional Program Coordinator of the United Church of Christ (UCC) Commission for Racial Justice.

His civil rights activism continued throughout the 1970s after he was sent to Wilmington, North Carolina, to desegregate the public school system. He was arrested and convicted on charges of arson and conspiracy as part of the "Wilmington 10" in 1972 after a local grocery store was firebombed. He was sentenced to 34 years for the crime and spent nearly a decade in prison, during which time he wrote two books. The charges were proved false, and he was released after a U.S. district court ruling in 1980. While in prison, Chavis was escorted to Duke University's divinity school to attend classes as part of a study-release program, obtaining a master of divinity degree there and a PhD in ministry from Howard University in 1981.

In 1983, Chavis went back to work for the UCC, where he was named executive director in 1985. He is credited with being one of the people who coined the term *environmental racism* after he participated in the protests over a hazardous waste dump in Warren County, North Carolina, where the population was 75 percent black. The landfill was constructed, but the protesters successfully stopped the state's plan to put another landfill and an incinerator there. As head of the UCC's Commission for Racial Justice, he was one of the authors of the landmark report on toxic waste siting that showed landfills and other facilities were predominantly located in communities of color. The first National People of Color Environmental Leadership Summit was held in 1987, where Chavis made an address that focused on environmental devastation facing minority communities. In 1993, he became the youngest person ever named as director of the National Association for the Advancement of Colored People (NAACP), where he served until he was fired by the organization's board of directors in 1994 over claims he diverted funds for a sexual harassment case in which he was involved. He converted to the Nation of Islam in 1997, becoming Benjamin Chavis Muhammad, and was ousted as a minister by the UCC. In 2001, as Minister Benjamin Muhammad, he cofounded the Hip-Hop Action Summit, which is dedicated to fighting the war on poverty and injustice.

Yvon Chouinard (b. 1938)

Although the founder of Patagonia calls his outdoor clothing and equipment company "an experiment," Yvon Chouinard has expanded his business to the point where he gives some of it away—1 percent of his gross revenues, to be exact. Known for recycling discarded plastic bottles and turning them into fleece jackets, he is also an expert rock and mountain climber, writer, and surfer. His 2005 book, *Let My People Go Surfing,* is part autobiography and part business guide.

He was born in Lewiston, Maine, the son of a blacksmith father. His family moved to Burbank, California, in 1946, and he joined the Sierra Club and founded the Southern California Falconry Club in 1953. In order to save money, he began forging his own climbing equipment in 1957, and he started Chouinard Equipment for Alpinists, selling gear out of the back of his car. In between adventures, he spent two years at a community college and two years in the military in Korea. In 1972, he began selling rugby shirts through a new company he founded, Patagonia, which has become a leader in both enhancing the workplace environment and protecting the natural environment. In 1993, the recycled plastic fleece products were added to the company's catalogue, and more recently, the efforts have expanded to include unusable second-quality fabrics and used clothing, becoming the world's first garment recycling firm, called Common Threads, in 2005. Recycling polyester clothing, Chouinard says, uses 76 percent less energy than if the product were made out of virgin petroleum.

Although the recycled soda bottle fleece is a company staple, Patagonia is working to "green" the company at every level. It became the first California company to use renewable energy to power all its buildings. Patagonia was also the first textile company in the world to print its commercial catalogue on recycled paper. Chouinard's son Fletcher is crafting a line of surfboards made of nontoxic materials, and the elder Chouinard has a new vision, called the Ocean Initiative. In 2001, he created One Percent for the Planet, an alliance of 500 businesses that pledge to donate 1 percent of their gross revenues to environmental protection. For Patagonia, the tithe now amounts to about $2.7 million per year.

Barry Commoner (b. 1917)

Barry Commoner was among the first scientists to become active in the nascent environmental movement, long before many of the environmental problems such as air and water pollution, nuclear waste, and toxins made headlines. He was born in New York City, and like many who later joined the movement, credits his family with encouraging his interests in science when he was given a microscope. He attended Columbia College, receiving a degree in zoology in 1937, and then received a master's degree and a PhD in biology from Harvard, where he completed his graduate work in 1941. He taught at Queen's College in New York City and was drafted by the Naval Reserve.

After being discharged, he worked as an associate editor at the magazine *Science Illustrated* and then taught at Washington University in St. Louis, where he became chair of the department of botany in 1965. He received a federal grant to set up the Center for the Biology of Natural Systems (CBNS) at the university, creating one of the first interdisciplinary programs to look at environmental problems. In 1970, at the height of the environmental movement, *Time* magazine called him "the Paul Revere of Ecology" because of his early warnings about pollution and the need for finding alternative sources of energy.

In the early 1980s, Commoner began studying the role of dioxins that were released when trash was burned, and he became an outspoken opponent of incineration projects when the industry was experiencing rapid growth in the United States. He ran for president of the United States as the representative of the Citizens Party in 1980, receiving 235,000 popular votes in the election, in which Ronald Reagan was chosen as president. He returned to Queens College in 1981, as did the CBNS. Although Commoner retired from teaching in 1987, he remained on the college's staff as director of the Center until 2000 and is now its director emeritus and a senior scientist.

Paul Connett (b. 1940)

Paul Connett's name is not well known outside the circle of anti-incineration activists and academicians who are familiar with his role in opposing incineration facilities in the Northeast. He deserves more credit, many have argued, for bringing science to policy debates over these issues. He was born in Sussex, England, and graduated from Cambridge University in 1962, majoring in natural science. He taught school for four years and became active in a group he and his wife, Ellen, founded, Operation Omega, to oppose the war in Bangladesh. They left the United Kingdom in 1979 so Paul could attend Dartmouth College, where he received his PhD in chemistry in 1983.

They moved to Canton, New York, where he worked at St. Lawrence University. From friends and colleagues, he heard about a proposed municipal waste incineration facility in Ogdensburg, about 17 miles from their home. Although he initially thought the project had some clear benefits because of the energy that would be produced and the level of government regulation that would be applied, his research on the plant's operations led to a turning point in his life in 1985. Connett was able to use his skills as a researcher and chemist to investigate some of the issues citizens were questioning on several projects that were being proposed. In one community, the issue of noise from a facility was being used by opponents, and he brought in information about the health risks from exposure to dioxins. He began to work with Barry Commoner in taking on industry representatives at public forums, and they formed the National Coalition against Mass Burn Incineration and for Safe Alternatives in October 1985. He also became part of the Citizens Conferences on Dioxins. In April 1988, his wife took on responsibility for producing a newsletter, *Waste Not,* and they helped produce a series of videotapes that presented the evidence against incineration that were distributed nationwide. He and Ellen were named the Conservationists of the Year from the Environmental Planning Lobby in 1990, and Paul received a Leadership Award from the National Campaign against the Misuse of Pesticides in 2000. He has shifted his interests to studying the effects of fluoride as part of the Fluoride Action Network. He is now an emeritus professor at St. Lawrence University, where his initial anti-incineration interests began.

Lois Gibbs (b. 1951)

Lois Gibbs's name is synonymous with community activism, although most people are unaware that she started out as a shy housewife before she became embroiled in the battle to clean up her neighborhood near Niagara Falls, New York. She was born in Buffalo, New York, and married a chemical worker; they moved to a new development called Love Canal. It was built on the site of the old Hooker Chemical Company, which had been covered with dirt in 1953 to be used for home sites and a school. A local reporter had written a series of articles in 1978 about chemicals that were found leaking from the ground there, and when her son began experiencing various illnesses, Gibbs started contacting her neighbors to see if they were concerned about any health problems that might be connected to the contamination.

After a series of public hearings by the New York State Department of Public Health, an elementary school that had been built at Love Canal in 1955 was closed in 1978. That same year, federal officials declared the site to be an emergency area, and the governor signed an order to relocate 239 families, many of whom had experienced miscarriages and children with birth defects.

Gibbs organized local homeowners who had not been relocated, seeking to have the government purchase their houses at a fair market value. State officials stalled on implementing the governor's relocation orders, and in 1980, angry residents protested and held two EPA officials hostage until the government agreed to relocate 810 families. Despite a $15 million federal authorization to buy their houses, some residents decided to stay.

Gibbs and her family moved to Washington, D.C., where she founded the Citizen's Clearinghouse for Hazardous Waste. The group was renamed the Center for Health, Environment and Justice in 1997 and moved to Falls Church, Virginia. With more than 27,000 members, it is one of the most prestigious coalitions of grassroots-level hazardous waste groups in the world.

Pete Grogan (b. 1949)

Pete Grogan is an example of a baby boomer who is convinced that it is possible to work at both the grassroots level and within a large corporation to deal with the issues of waste. He was born in Newark, New Jersey, and says his interest in trash grew out of his parents' encouragement of recycling and, as a child, seeing bits of incinerated garbage on the beach. He graduated with a bachelor's degree in psychology from the University of Colorado, Boulder, in 1974. He began working with juvenile offenders and started a recycling program to help pay for sports equipment, and he and a friend started a grassroots recycling effort, EcoCycle, in 1976. Volunteers collected recyclables and then sold what they could to reprocessors. His university contacts helped the group get student volunteers. The organization also received funding from the city of Boulder, the county, and the EPA.

Grogan left EcoCycle in 1987 to serve as an international consultant with R. W. Beck and Associates, assisting companies and government agencies in both the United States and abroad seeking to develop their own recycling and sold waste planning programs. He also began working with industry groups to increase the percentage of their waste that could be recycled. In 1994, he joined Weyerhauser and is now manager of Market Development for the company's recycling division. The company collects and sorts 50 grades of recovered paper at its U.S. facilities. In 2006, in recognition of his 30 years of effort to promote recycling, Grogan was given the Lifetime Achievement Award and Recycler of the Year award from the National Recycling Coalition.

Kenny Guinn (b. 1936)

Kenny Guinn held no elected office prior to becoming the Republican governor of Nevada in 1998. After serving two terms, he stepped down from the state's highest office in 2007, but he remains active as one of the most powerful, and strident, opponents to the Yucca Mountain nuclear waste depository in his state. He is described as a RINO, or "Republican in Name Only," because of his moderate stance on most issues and his appeal to Democratic voters. He was named one of the nation's five best governors by *Time* magazine in 2005.

Guinn was born in Garland, Arkansas, but grew up in the central California town of Exeter. He received his bachelor's degree in physical education from Fresno State University in 1957, followed by a doctorate in physical education from Utah State University in 1970. He and his family moved to Las Vegas in 1964, where he served as the superintendent of education for the Clark County School District from 1969 to 1978. He next served as vice president of Nevada Savings and Loan (NSL) from 1978 to 1987; he was named president and chairman of the board when the organization changed its name to PriMerit Bank. From that position, Guinn moved on to serve as chairman and chief executive officer of Southwest Gas Corporation and, from 1993 to 1997, was chairman of the board of directors of Southwest Gas. He briefly served as interim president of the University of Nevada, Las Vegas, from 1994 to 1995.

Guinn was an early opponent of the Yucca Mountain facility, which was designed to hold as many as 77,000 tons of highly radioactive material at a site about 90 miles northwest of Las Vegas. He led other state officials and the Nevada congressional delegation in questioning the suitability of the site and the potential for groundwater contamination there. Fears about the durability of waste containers, interstate transport, and geologic safety united Guinn with environmental groups and activists and with President Bill Clinton, who also opposed the plan for the storage site at the edge of the former Nevada Nuclear Test Site. In April 2002, Guinn used his veto power under special congressional rules and the Nuclear Waste Policy Act to reject the U.S. Department of

Energy's recommendation that Yucca Mountain be the nation's long-term depository for nuclear waste. The Republican-controlled House of Representatives and Senate later voted to override Guinn's veto, paving the way for the Yucca Mountain project to move forward. President George W. Bush approved the site in 2004. Guinn countered by taking the federal government to court, contending that Nevada, which has no nuclear plants of its own, was bearing an unfair burden. Guinn also challenged EPA and Nuclear Regulatory Commission rules related to the plant's licensing and operation, a process that could take at least five more years. When he left office in January 2007, Guinn's postgubernatorial life shifted considerably; he is now a member of the board of directors for the MGM Mirage Hotel in Las Vegas.

Alice Hamilton (1869–1970)

Alice Hamilton grew up as one of five children in New York City, where she was born into a family that was well off financially. She attended the private and prestigious Miss Porter's School and decided to enter the field of medicine, an occupation that would allow her to work almost anywhere. She attended the Fort Wayne College of Medicine and the University of Michigan, and although she did not complete an undergraduate degree, she received her medical degree in 1893. She had extensive training at several hospitals and laboratories in both the United States and Europe, and after deciding she wanted to be a researcher rather than a physician, she accepted a position as professor of pathology at Northwestern University's Women's Medical School in 1897. She made a study of the typhoid epidemic that was raging in Chicago and determined that the disease was being spread by flies, one of many discoveries that would lead her to investigate workplace conditions and the role of chemicals and toxic waste. She began a lifelong effort to "treat the excesses of industrialization."

In 1911, Hamilton accepted a position as a special investigator for the new U.S. Department of Labor, which required her to do field work at many high-risk work sites. Over the years, she would investigate the steel and high-explosives industries, where she found the chemicals used and the failure to deal with industrial waste properly were responsible for many deaths. She became Harvard University's first female faculty member in 1919 in the industrial medicine program, although she was subjected to widespread gender discrimination. Her 1925 book, *Industrial Poisons in the United States,* was the first textbook on the topic at a time when little attention was being paid to industrial practices and what would later be known as industrial waste. Another book, *Exploring the Dangerous Trades,* published in 1943, marked the end of her work on toxins in the workplace, but it did not end her social activism. She fought against child labor, supported the League of Nations and the Equal Rights Amendment, became a pacifist, and opposed the war in Vietnam. She died at her home in Connecticut at age 101 and was commemorated by a U.S. Postal Service stamp in 1995.

H. Wayne Huizenga (b. 1937)

A flamboyant entrepreneur, billionaire H. Wayne Huizenga has an important place in the history of waste management, having built three Fortune 500 corporations. Despite his financial successes, critics argue that his alleged ties to organized crime and unfair business practices have given the industry a black eye that refuses to go away.

He was born in Evergreen Park, Illinois, and spent part of his youth in Chicago until his family moved to Florida in 1953, where he graduated from Pine Crest High School in Ft. Lauderdale. He moved back to Chicago and in 1956 entered Calvin College in Grand Rapids, Michigan, but left after three semesters. He enlisted in the U.S. Army reserve in 1959, and in 1960 returned with his wife to Ft. Lauderdale. He started working for a local garbage hauler and in 1962 started his own garbage company, Southern Sanitation Service. In 1970, he merged his business with Dean Buntrock's Chicago garbage company, forming Waste Management, Inc. He stepped down as vice chairman of Waste Management at age 46 to move on to other business interests, but in 1995, he purchased Republic Industries, which grew to be the third-largest waste management company in the United States.

Although waste management was the initial source of his wealth, Huizenga expanded his business interests in the 1980s and 1990s. He bought out the Blockbuster Video chain of stores for $18 million, and he began buying up small video stores, taking the company public in 1989. He also started a hotel chain, Extended Stay America, and started the first nationwide auto dealer, AutoNation. In 1990, he bought an interest in the National

Football League's Miami Dolphins, and in 1991, he brought the Florida Marlins, a Major League Baseball expansion team, to Miami, followed by the National Hockey League's Florida Panthers in 1992.

Huizenga has a controversial public persona, which was ignited in 1994, when a newspaper article made allegations about his assault of a sales prospect who refused to do business with him, and stories arose about spousal abuse, ties with organized crime, illegal political contributions, unfair competition practices, and a disregard of environmental laws. Despite the charges, he has been named *Financial World*'s CEO of the Year; has been named Ernst & Young's U.S. Entrepreneur of the Year; and in 2007, joined presidential candidate Mitt Romney's Florida statewide finance committee. As a noted Florida philanthropist, he has supported many projects, including the Huizenga Business School at Nova Southeastern University in Ft. Lauderdale.

Mary McDowell (1854–1936)

"Fighting Mary" McDowell is also known as "The Angel of the Stockyards." She was born in Cincinnati and raised in Chicago, the daughter of a steel-rolling mill owner who had also served in the Union Army during the Civil War. Her history of social activism began with the Methodist Church, and in the 1880s, when her family moved to Evanston, Illinois, she joined the Young Women's Christian Temperance Union. The group was starting the kindergarten movement for young children, and McDowell became a teacher in New York City for a short time. Her work with the University of Chicago Settlement House helped those living near the stockyards and immigrant families. McDowell lived there for the rest of her life; it was renamed the Mary McDowell Settlement in her honor.

McDowell became active in garbage-reform activities as part of her work in the Packingtown neighborhood adjacent to Chicago's open pit dumps. Sanitation was almost nonexistent in Packingtown, so she organized various cleanup campaigns. She also worked to reform the municipal garbage collection system, which was controlled by powerful political officials. In 1911, she toured Europe to see the various sanitation and waste management systems being used there. When she returned, she started a public campaign to organize garbage committees using women's clubs as the base of membership. The city's first garbage reduction plant was built in 1913, although McDowell did not live to see a modern waste treatment facility. The remainder of her life was spent in trying to improve race relations, the inspection of packing plants, and the building of city parks and community gardens.

Martin Melosi (b. 1947)

One of the most prolific historians and authors on the subject of trash, landfills, urban sanitation, and waste management, Martin Melosi has a resume 23 pages long and combines academic research with public service and litigation support. He was born in San Jose, California, and received a bachelor's degree in 1969 and a master's degree in 1971 from the University of Montana, where he majored in history. In 1972, he received the George P. Hammond Prize for the best paper written by a graduate student from Phi Alpha Theta, the national history honorary society. He went on to receive a PhD in history from the University of Texas at Houston, followed by a series of academic appointments in Texas; Paris; Helsinki, Finland; and Odense, Denmark. He is now the Distinguished University Professor of History at the University of Houston, where he has taught since 1984.

Studying garbage represents a unique academic niche, and Melosi has researched the problem from every conceivable perspective, with a brief stint as a biographer of Thomas Edison and author of a study of the attack on Pearl Harbor. His 1980 book, *Pollution and Reform in American Cities, 1870–1930,* is considered to be one of the finest historical treatments of early waste management, followed in 1981 by *Garbage in the Cities: Refuse, Reform, and the Environment 1880–1980.* The two books provide an outstanding overview of how the United States transformed its waste practices as it changed from an agrarian to an industrial nation, and the role of issues such as recycling during the development of the environmental movement.

Melosi has served as president of several national organizations, including the American Society for Environmental History, the National Council on Public History, and the Public Works Historical Society. Besides being a prolific author, Melosi is known for public service activities, ranging from serving as a consultant to the Tempe, Arizona, Historical Museum and as a member of the Advisory Board of the Environmental Institute of Houston to being an exhibit adviser to the Smithsonian Institution and serving as a film consultant. He has also worked as an expert witness in litigation involving Shell Oil Company.

Penny Newman (b. 1947)

In the 1980s, Penny Newman was called a superwoman for her activism at the Stringfellow Acid Pits, California's top-priority Superfund site. As the West Coast organizer for the Citizen's Clearinghouse for Hazardous Waste, founded by Love Canal activist Lois Gibbs, Newman would become one of the waste industry's most feared antagonists.

Her life has focused on the rural California community of Glen Avon, about 60 miles east of Los Angeles. Newman attended Riverside City College and California State University, Fullerton, where she received a bachelor's degree in communicative disorders in 1981. She worked as an instructional aide for a local school district and as a substitute teacher, later working as a special education teacher for children with severe language disorders.

In 1969, heavy rains caused a local earthen dam to overflow, carrying waste from a toxic dump into the small community where she and her family lived. Her son began having bouts of asthma-related breathing problems and dizzy spells, as did many other residents. Newman became one of the leaders of Concerned Neighbors in Action (CNA), a group formed in 1979 to protest the lack of federal attention to the Stringfellow Acid Pits, which had opened in 1956 as a dump for more than 34 million gallons of hazardous liquid waste. Newman became a full-time employee at CNA in 1986. In 1987, the group brought suit before the U.S. Supreme Court to try to force the government to clean up the site, a process that has taken nearly two decades of Newman's life.

Newman is now the executive director for the Center for Community Action and Environmental Justice, a nonprofit organization. Her honors include awards from the South Coast Air Quality Management District, the California Senate, U.S. Senator Barbara Boxer, the Environmental Protection Agency, and the American Public Health Association.

Paul Palmer (b. 1938)

The idea of "zero waste" has joined the lexicon of industrial designers, activists, and policy makers. The man credited with first using the phrase *zero waste* publicly is Paul Palmer, whose name is much less familiar in waste management history. The concept has been interpreted in numerous ways, but it is generally defined as changing and redesigning the production of goods to consider the entire life cycle of resources. It is endorsed by state legislators and city councils and embraced by state waste managers and entire communities. The term is associated with words and phrases such as *vision, design principle, philosophy, systems approach, producer responsibility,* and *true cost accounting.*

Palmer was born in the Bronx, New York, in 1938, and was awarded his bachelor of science degree in chemistry at Queens College in New York City in 1959. He went on to receive a PhD in physical chemistry from Yale University in 1966. He took a job teaching chemistry at the Middle East Technical University in Ankara, Turkey, for two years, then taught for a year as a visiting scientist at Oersted Institute in Copenhagen, Denmark. When he returned to the United States, Palmer served as a lecturer in the chemistry department of Indiana University in Bloomington. His relatively placid academic career changed when he moved to California and founded Zero Waste Systems (ZWS), Inc., in 1973 in Berkeley. The company initially sought to find new uses for chemicals that were left over from the electronics industry, offering to take them for free. New and usable laboratory chemicals were added to ZWS's huge inventory and then sold at half price to scientists, companies, and those conducting experiments. ZWS also collected developer / rinse solvent produced by various companies, which was put into small cans as lacquer thinner, and reflow oil created by printed circuit makers that was filtered and remarketed for use in oil well production. The company was termed an "active waste exchange" by the EPA.

Palmer's book *Getting to Zero Waste,* published in 2005, explains how in his ideal model, individuals and businesses would have to find creative ways to deal with waste, rather than relying on garbage haulers and dumps. Consumers could go to large facilities, where they could refill containers or bottles with the items they needed. If manufacturers are held responsible for the waste they create, he expects they would change the way their products are manufactured. He established the Zero Waste Institute in Sebastopol, California, and continues to try to convince others of the need to use resources more intelligently.

William Rathje (b. 1945)

Imagine being called "The Indiana Jones of Solid Waste." That is the informal title that has been bestowed on William Rathje, emeritus professor of archaeology from the University of Arizona, who is better known as the founding director of the Garbage Project.

Born in South Bend, Indiana, he grew up in Wheaton, Illinois, the son of a teacher and an air traffic controller. At age 11, his grandparents gave him a copy of the book *The Wonderful World of Archaeology,* a defining experience in his young life. When he left Wheaton in 1963 to study archaeology at the University of Arizona in Tucson, he decided to focus his attention on Mayan culture, and after he received his bachelor's degree in 1967, he went to Harvard, where he was awarded a PhD in 1971. He returned to Tucson to teach at his alma mater and continued to study Mayan society by excavating various burial sites. In the process, he developed a theory that during the Late Classic period, Mayan society became more aristocratic and developed particular hereditary groups. He also theorized that prehistoric burial practices were linked to contemporary ones in how the dead are treated, and he studied trading zones in prehistoric societies.

The Garbage Project had its beginnings in 1971, when Rathje taught an anthropology class at the University of Arizona that was designed to teach archaeological methods. Students undertook independent projects that were to show the links between various kinds of artifacts and various kinds of behavior. Two of the students explored the relationship between mental stereotypes and physical realities by collecting garbage from two households in an affluent part of Tucson, and two others in a poor part of town. Their goal was to determine whether garbage samples could be used to gauge behavior. The following year, students began "borrowing" and sorting household garbage from different parts of Tucson; in 1973, when the Garbage Project became more structured, the city's sanitation division delivered randomly chosen household pickups to an analysis site at a maintenance yard. The operations were shifted to a site where the university's dumpsters were parked in 1984, across from the school's stadium. Later, the Garbage Project shifted its base of operations to municipal landfills for excavation.

Florence Robinson (b. 1937)

Florence Robinson, who was born in northern Louisiana, has been called a "reluctant warrior," and although she is not nearly as well known as Love Canal activist Lois Gibbs, she and Gibbs share two things in common. Both have been active as "ordinary citizens" in trying to focus attention on the problems of hazardous waste, and together, they shared the 1998 Heinz Award for the Environment.

Robinson lived in the small community of Alsen, Louisiana, near Devil's Swamp, where industrial waste was routinely dumped in a pit opened there in 1964. Alsen is also home to 11 petrochemical plants. She attended Southern University, majoring in biology, and later returned to teach there in the 1970s. In 1981, Alsen's residents, who are mostly poor and black, filed suit against Rollins Environmental Services, one of the worst polluters in the area. The company decided to expand in 1986, and Robinson and her neighbors held demonstrations in protest. She purchased a computer so she could track emissions from industrial plants through the federal Toxic Release Inventory. She began testifying at various hearings on at-risk communities near waste sites and joined activists who had developed the "People's Agenda on Superfund" in 1993 and 1995. In 1997, her efforts convinced one of the area's major polluters to stop burning hazardous waste in an incinerator, and the facility was dismantled a year later. She helped found the North Baton Rouge Environmental Association, and in 1999, she joined activists in protesting the incineration of napalm; in 2001, she and her students participated in the Greenpeace Celebrity Tour of Cancer Alley, the 85-mile strip of land along the Mississippi River in Louisiana.

Her anti–toxic waste activism has come at a price; she has been diagnosed with chronic fatigue syndrome and anxiety, and she has lost her sense of smell. In 1998, she left Alsen for Baton Rouge after she said she had difficulty breathing and worried about the health of her dogs. She believed that her exposure to the petrochemical companies' toxic fumes and waste had damaged her health, although she did not sell her home because she said she did not want anyone else exposed to the risks.

Elizabeth Royte (b. 1960)

The idea of spending the weekend reading about garbage may not sound appealing to most people, but if the author being read is Elizabeth Royte, the time is likely well spent. Her book *Garbage Land: The Secret Trail of Trash* was named one of the *New York Times*'s 100 Notable Books of 2005. Royte's style of investigative journalism appeals to those who want information and a witty writing style, such as her notation that maggots found in garbage are known as "disco rice" by sanitation workers.

She was born in Boston and says she spent a lot of time outdoors when she was growing up. She graduated from Bard College in 1981 and started her writing career as an intern at *The Nation*. At the time, she hoped to get a job as an editor, but her freelance writing gradually developed into a science/environmental journalism niche. In 1999, she was named an Alicia Patterson Foundation fellow and spent a year of funded research at a biological research station. Now based in New York City, she has written several books and contributes to *Harper's, Outside, National Geographic*, the *New York Times Magazine*, and *Life*.

Although she has written about numerous environmental topics, her work on waste is unique among nonfiction books. In one article, Royte follows her own household garbage to see where it goes after it leaves her home and is surprised at how much organic waste she and her family produce. In a subsequent article, she writes about the waste and debris from Hurricane Katrina, and in Huffington Post blogs, she has dealt with topics such as paper versus plastic and compost in ways that have given new meaning to "talkin' trash."

William Sanjour (b. 1933)

Now retired, William Sanjour was one of the best-known career staff members of the U.S. Environmental Protection Agency because of his role as an agency whistle-blower. In 1979, he was transferred from a position of influence at the EPA to one where his position had no assigned duties. He has been the subject of harassment and is considered by many to be the driving force behind the 1976 Resource Conservation and Recovery Act (RCRA).

He was born in New York City and served in the U.S. Army before attending the City College of New York, where he majored in physics and graduated with a bachelor's degree in 1958, followed by a master's degree in physics from Columbia University in 1960. One of his first jobs was as an operations analyst for the U.S. Navy at its think tank, the Center for Naval Analysis. From there he worked for the American Cyanamid Company from 1964 to 1966, and then in 1967 as a management consultant for Ernst and Ernst. His job was to develop computer simulations for reducing air pollutants, and from there he became a consultant for the EPA from 1972 to 1974, when he joined the agency full time.

He accepted a position with the EPA's Hazardous Waste Division, supervising studies to assess the scope of the problem and how sites might be cleaned up. The studies conducted under his leadership were sufficient to convince Congress to enact RCRA at a time when the environmental movement was combining advocacy with scientific analysis. He helped draft the rules that would eventually implement the provisions of RCRA, although little support for enforcement was provided under the administration of President Jimmy Carter. His efforts to publicize the EPA's lack of compliance with the law resulted in his transfer to a position with no duties. He fought the transfer and, after a year-long legal battle, won appointment as head of the Hazardous Waste Implementation Branch while continuing to expose fraud and abuse within the EPA. In his nonworking hours, he began assisting grassroots environmental groups and testified numerous times before Congress.

In 1984, he voluntarily accepted a transfer to the Office of Technology Assessment and wrote a critical report on the EPA's Superfund efforts, an action that cost him his position as branch chief when he returned to the EPA. Reassigned as a policy analyst, he fought efforts to silence whistle-blowers and was successful in a landmark court decision on his complaints about treatment at the agency. He retired in 2001, although he continues his activism as a member of the North Carolina Waste Awareness and Reduction Network and the National Whistleblower Center.

Wilma Subra (b. 1943)

Although some call her "St. Wilma" and one writer called her a "genial grandmother," Wilma Subra, a native of Louisiana, has worked tirelessly, and in some cases has taken on major corporations as "a top gun for the environmental movement," to help the residents of her state. She has been rewarded not only with international awards but also with the sense of accomplishment that typifies her upbringing in the small town of Morgan City. During the summers while she was in high school, she worked for her father, who ran a small laboratory grinding oyster shells, which she says was the source of her interest in chemistry. She attended the University of Southwestern Louisiana in Lafayette, where she earned a bachelor's degree in 1965 and a master's degree in 1966 in chemistry and microbiology. She joined the Gulf South Research Institute in 1967 as a microbiologist and biostatistician, where she worked for 14 years. At Gulf South, she and her colleagues conducted toxicological studies and projects with the National Institute of Cancer Research at a time when the field was just developing. They also worked on quick-response projects for the EPA to study why a high rate of disease was occurring in certain communities, including New York's Love Canal.

Her top-gun reputation comes from working with more than 800 communities over the last 25 years, assisting residents who are concerned about toxic waste contamination in their neighborhoods. She has focused her attention on Cancer Alley because of the illnesses reported there.

Most of her work is done for free, paid for by the chemical consulting work she does for food companies and other clients as president of the Subra Company in New Iberia, Louisiana. In 1999, she was awarded a $370,000 MacArthur Foundation "genius grant" that helps fund her activities. The award specifically mentioned her work in Morgan City against the Marine Shale Facility, which was a hazardous waste incinerator that claimed to be a recycler and was finally shut down in 1996.

Terri Swearingen (b. 1956)

Terri Swearingen was born in the small, rural Appalachian community of East Liverpool, Ohio, the daughter of a steelworker father and a mother who worked in a school cafeteria. In 1978, she received a nursing degree from the Ohio Valley Hospital School of Nursing and also was trained as a dental technician. She worked in her husband's dental office, where in 1982, a patient mentioned that the world's largest hazardous waste incinerator was being planned for the town. Waste Technologies Industries, which began its plans in 1980, was to be built within 1,100 yards of an elementary school in a low-income minority neighborhood and 320 feet from a nearby house.

Swearingen and her neighbors began writing letters in opposition to the incinerator and believed that the EPA would never grant a permit to the company, which was slated to be built on the banks of the Ohio River. But construction began in 1989, and a year later, Swearingen committed herself full time to fight the plant's operation, founding a grassroots group, the Tri-State Environmental Council. She believed that the siting of the facility was an example of environmental racism, because the community in which it was located was poor and had been chosen for that reason. Of the 13,000 residents of East Liverpool, 500 were black, and all of them lived in the vicinity of the incinerator.

In 1997, she was named a winner of the Goldman Environmental Prize for North America, and she also received the 1999 William E. Gibson Award from the General Assembly of the Presbyterian Church. Swearingen has been named a Women at Their Best winner by Cover Girl and *Glamour* magazine and was named one of six "ecowarriors" by *marie claire* magazine in 1997. In 1994, she was named one of the 50 most promising leaders in the United States by *Time* magazine, and she is a member of the Citizens Clearinghouse for Hazardous Waste Grassroots Hall of Fame.

Mike Synar (1950–1996)

Rep. Mike Synar of Oklahoma, who served 16 years in Congress, was known for taking on political challenges, whether the foe involved the tobacco industry, the gun lobby, Oklahoma's energy interests, or his own Democratic party. In 1987, he took on the U.S. Department of Defense (DOD) in a frustrating attempt to find out how the military was dealing with hazardous waste.

Michael Lynn Synar was born in Vanity, Oklahoma, attended local schools in Muskogee, and had been named the nation's top

4-H club member; his family had been named the "outstanding family in the U.S." by the All-American Family Institute. He attended the University of Oklahoma, Norman, where he was awarded a bachelor's degree in 1972, and was named a Rotary International Scholar, attending the University of Edinburgh Graduate School of Economics, in Scotland, in 1973. His master's degree was completed in 1974 at Northwestern University, and he returned to the University of Oklahoma to attend law school, which he completed in 1977. He ran for the U.S. House of Representatives in 1978 and served there until he lost reelection in 1994.

Synar served as chair of the congressional Subcommittee on Environment, Energy and Natural Resources in 1987, and in November that year held hearings on the efforts of the DOD to address its environmental problems. The issue had been one of special interest to him, and one on which he staked his political career. Tinker Air Force Base was in his home state, and his staff had conducted considerable research in the four years since the military's previous report before Congress. An official from the General Accounting Office (now the Government Accountability Office) testified that these investigations had found that the military had problems with waste storage and disposal, record keeping and tracking of hazardous waste shipments, hazardous waste container management, and ground water management, with numerous environmental violations the previous year. Synar's tenacious questioning of Defense officials and his prodding of the Pentagon to clean up military installations was one factor leading to his receipt of the Profiles in Courage award from the John F. Kennedy Library Foundation on May 8, 1995. Synar developed a reputation as one of the most knowledgeable members of Congress on the issue of the military's waste remediation efforts. He also was active in other waste management issues, introducing legislation to amend the Solid Waste Disposal Act to ensure that any waste exported from the United States to foreign countries is managed to protect human health and the environment. Synar died of brain cancer in Washington, D.C., in January 1996.

Will Toor (b. 1961)

As a native of Pittsburgh, Will Toor grew up in a highly charged academic world, the son of two university professors. At the age of 17, he graduated from Carnegie Mellon University with a bachelor of science degree in physics. He treated himself to a cross-country hitchhiking tour and ended up in Boulder, Colorado, in 1980, where he decided to stay. Boulder is also the city where his early work at a recycling center led to a political career that has allowed him to expand his initial agenda to one encompassing local sustainability.

As an activist against nuclear weapons testing at the nearby Rocky Flats facility, he became friends with the area's environmental leaders and joined the community nonprofit recycling group EcoCycle, founded in 1976. He began by working on Saturdays riding around in the back of an old school bus helping pick up recyclables, then working full time in the recycling yard. Toor helped develop Boulder's curbside recycling program, turning it into one that would be a model for the rest of the United States. It is now the largest nonprofit recycler in the nation.

He decided to go back to school and enrolled at the University of Chicago, receiving a master's degree in physics in 1987 and a PhD in 1992. As a graduate student, he coordinated the university's recycling program and then returned to Boulder in 1992 as director of the University of Colorado's Environmental Center, where he served until 2004. He was also an instructor in the university's environmental studies program, teaching courses on environmental leadership and transportation policy. Toor's leadership of Boulder's recycling programs has been the subject of several awards, as has his work with the Environmental Center. The EPA honored the university with awards for recycling, green power, and its student bus program, and the National Recycling Coalition awarded the university with its Outstanding School Recycling Program award. His political career started in 1997, when he ran successfully for the Boulder City Council, and in 1998, he became mayor. Toor was elected to the Boulder County Board of Commissioners in 2004, and has spearheaded the county's sustainability initiative.

Jean Lacey Vincenz (1894–1989)

Jean Lacey Vincenz was born in Enfield, Illinois, in 1894, and attended junior college in Fresno, California. He received a bachelor's degree in civil engineering from Stanford University in 1918, and over a 10-year period served as commissioner of public works, city engineer, and manager of public utilities in Fresno.

While this might not seem especially notable, Vincenz's life is remarkable because he was responsible for the development of the United States' first sanitary landfill.

As commissioner of public works for Fresno, he recommended that the city not renew an existing franchise contract with the Fresno Disposal Company, which operated an incinerator for burning trash. He had studied the British system of controlled tipping techniques, visited other California cities to see how they dealt with municipal waste, and talked to engineers about how best to operate a landfill. Vincenz believed that different strategies needed to be implemented to build trenches, compact the waste, and cover it with a deeper layer of dirt than was normally used elsewhere. No one had ever placed an emphasis on compaction, and he believed this would create a landfill that was more efficient. He helped develop a waste system for Fresno's trash and opened the city's sanitary landfill in 1937.

In 1941, he joined the Army Corps of Engineers Repairs and Utilities Division as assistant chief in Washington, D.C., where he worked until 1947. He helped the military adopt the sanitary landfill as a standard disposal method after World War II. He then was selected as the City of San Diego's Public Works director, a position he held until 1962. His honors included being named the president of the American Public Works Association in 1960.

George E. Waring Jr. (1833–1898)

George E. Waring Jr., a Civil War veteran known as "The Colonel," began work as an engineer, designing and constructing municipal sewer systems. When the leaders of the civic reform movement were able to defeat New York City's Tammany Hall and elect William Strong as mayor, Waring was made commissioner of New York's City Street Cleaning Department, which was responsible for sweeping more than 400 miles of city streets and collecting trash from businesses and residences. Initially, he tried to work with the employees who were already part of the department, but he recognized almost immediately that he would have to reorganize it completely. He set up a model, military-style garbage management system in the United States, employing 2,000 "White Wings," as his workers were called, because they wore starched white uniforms. Each day, the workers participated in a morning roll call, and they were expected to follow an extensive set of written rules. His wife designed collection cans for picking up garbage, and the workers participated in city parades on holidays, marching in formation and accompanied by a band. He increased worker salaries, instituted an eight-hour work day, and established a formal grievance procedure.

Waring was among the first to set up a specialized system for handling and treating municipal waste, banning independent scavenging and cart haulers who went from door to door picking up refuse, and prohibiting city employees from sorting through garbage. Secondhand dealers had to be licensed to operate and worked under supervision, and residents had to presort their garbage in what would become the beginnings of modern recycling and source separation. Trash was not to be left at the curbside; residents put an elegant "call card" in their front window so that licensed handlers could retrieve it and haul it to one of the city's piers, where it was loaded onto barges. The barges were taken to Barren Island in Jamaica Bay, where the refuse was processed by Waring's new Sanitary Utilization Company. The "Apostle of Cleanliness" is credited with setting up the first comprehensive public-sector waste management and sewer systems in the United States. He served as commissioner of the city's Street Cleaning Department for three years and was ousted when Tammany Hall took back control of New York City in the 1898 elections.

Craig E. Williams (b. 1948)

The Goldman Environmental Prize is awarded each year to six grassroots environmental heroes representing each continent, and in 2006, the North American winner was Craig E. Williams, director of the U.S. Chemical Weapons Working Group (CWWG). Williams was given the prize for his work in guiding a citizens' coalition in opposition to the incineration of stockpiles of chemical weapons by the Pentagon. He founded the CWWG in 1990 as a coalition of grassroots groups on the regional and national levels who support the destruction of U.S. chemical weapons, but not through the process of incineration, which is more commonly used for burning municipal waste. The organization argues that in the United States, the incineration process is plagued with technical problems, lawsuits, growing citizen opposition, and gross mismanagement.

Williams was born in Flushing Meadows, Queens, New York, in 1948 and attended Central Florida Junior College. He went to the Defense Language Institute in Washington, D.C., in 1968, studying Vietnamese. After serving in the army in Vietnam from 1968 to 1969, he joined the steering committee of Vietnam Veterans against the War and then completed his education at Eastern Kentucky University, where he received a bachelor of arts degree in philosophy in 1978.

He became involved in the anti-incineration campaign in 1985, when he found out that the U.S. Department of Defense planned to build a facility at the Blue Grass Army Depot just eight miles from his home in Kentucky. Initially, he was a cofounder and member of the steering committee for Common Ground–Kentuckians for Safe Disposal of Chemical Weapons, and in 1990, he cofounded the Kentucky Environmental Foundation, the same year that he cofounded the CWWG. Williams and his family lived near the 6.2-mile radius around Blue Grass, known as the "immediate response zone," that would be under the greatest threat from a chemical accident there. He began attending scoping meetings required under the National Environmental Policy Act and designed to involve the public in the decision-making process. The army outpost is in Madison County, near Richmond, Kentucky, where 523 tons of chemical weapons were stored in underground bunkers.

In 1996, Williams claimed success when the funding for the Kentucky plant and another in Colorado was suspended. Williams has organized citizen summits on chemical weapons disposal, both in the United States and in Russia; served as a witness before congressional committees; and is an adviser on military toxic materials. His numerous commendations and awards range from groups such as the National Toxics Campaign Fund and Common Cause to the Sierra Club; Berea College; and the Union for Chemical Safety, Saratov, Russia.

6

Data and Documents

This chapter seeks to put the issue of waste management in context by providing some examples of the important documents, speeches, data, facts, and legislation that have shaped this critical environmental policy problem. The sections are arranged in both a topical and chronological manner and include both domestic and international perspectives. In some cases, the material provided is in the form of excerpts that explain key ideas and information. Sources are provided so that it is possible to examine the original material in more depth. Public documents, such as statutory material, are widely available in public libraries or on the Internet.

Early Efforts to Manage Waste

The 1970s are usually considered the period in which the United States made its most notable efforts to insure the protection of the environment. When Earth Day was first observed on April 22, 1970, Americans were beginning to feel a sense of urgency about cleaning up and protecting the nation's air, water, and public lands. Although President Richard M. Nixon is often credited with being "the environmental president" because of his administration's efforts to work with Congress to enact legislation for the protection of the environment, his predecessor, Lyndon B. Johnson, also has a noteworthy place in environmental history. In early 1968, he sent a special message to Congress that identified some of the nation's most compelling environmental problems and the need to solve them. Johnson's widely cited commentary on conservation, "To Renew a Nation," is noteworthy for two reasons. First, Johnson called for "The New Conservation" that would deal with the dangers of technology. Second, in addition to outlining the major challenges facing the country—water pollution control, safe community water supplies, oil pollution abatement, air pollution, noise control, surface mining, the protection of natural areas, and the development of marine sciences—he was among the first to focus on the issue of solid waste as one of the country's major issues. Congress would eventually heed Johnson's message with passage of the Resource Recovery Act of 1970, which shifted waste management away from simply disposing of waste and toward recycling and the reuse of materials, and a subsequent statute in 1976 that focused more attention on this problem. In these excerpts from his Message to the Congress from 1968, Johnson speaks vividly of the nation's conservation heritage and the need to manage its waste.

Excerpts from a Special Message to the Congress on Conservation, "To Renew a Nation," President Lyndon B. Johnson (March 8, 1968)

To the Congress of the United States:

Theodore Roosevelt made conservation more than a political issue in America. He made it a moral imperative. More than half a century ago, he sounded this warning:

"To skin and exhaust the land of using it so as to increase its usefulness, will result in undermining in the days of our children the very prosperity which we ought by right to hand down to them amplified and developed."

The conservation work that Roosevelt began was protection of our natural heritage for the enjoyment and enrichment of all the families of the land. That is work which never ends. It must be taken up anew by each succeeding generation, acting as trustees for the next.

But the conservation problems Theodore Roosevelt saw are dwarfed by new ones of our own day.

An unfolding technology has increased our economic strength and added to the convenience of our lives.

But that same technology—we know now—carries danger with it.

From the great smoke stacks of industry and from the exhausts of motors and machines, 130 million tons of soot, carbon and grime settle over the people and shroud the Nation's cities each year.

From towns, factories, and stockyards, wastes pollute our rivers and streams, endangering the waters we drink and use.

The debris of civilization litters the landscapes and spoils the beaches.

Conservation's concern now is not only for man's enjoyment—but for man's survival.

Fortunately, we have recognized the threat in time, and we have begun to meet it.

Through the landmark legislation of the past few years we are moving to bring a safe environment both to this generation, and to the America still unborn.

But the work of the new conservation, too—like the task we inherited from an earlier day—is unending. Technology is not something which happens once and then stands still. It grows and develops at an electric pace. And our efforts to keep it in harmony with human values must be intensified and accelerated. Indeed, technology itself is the tool with which these new environmental problems can be conquered.

In this Message I shall outline the steps which I believe America must take this year to preserve the natural heritage of its people—a broad heritage that must include not only the wilderness of the unbroken forest, but a safe environment for the crowded city.

In 1965, I recommended and the Congress approved a national planning, research and development program to find ways to dispose of the annual discard of solid wastes—millions of tons of garbage and rubbish, old automobile hulks, abandoned refrigerators, slaughterhouse refuse. This waste—enough to fill the Panama Canal four times over—mars the landscapes in cities, suburbia and countryside alike. It breeds disease-carrying insects and rodents, and much of it finds its way into the air and water.

The problem is not only to learn how to get rid of these substances—but also how to convert waste economically into useful materials. Millions of dollars of useful byproducts may go up in smoke, or be buried under the earth.

Already scientists working under the 1965 Act have learned much about how soils absorb and assimilate wastes. States and local communities have drawn up their plans for solid waste disposal. That Act expires in June, 1969.

To continue our efforts, I recommend a one-year extension of the Solid Waste Disposal Act.

In addition, I am directing the Director of the Office of Science and Technology working with the appropriate Cabinet officers to undertake a comprehensive review of current solid waste disposal technology. We want to find the solutions to two key problems:

How to bring down the present high cost of solid waste disposal.

How to improve and strengthen government-wide research and development in this field.

Three years ago, I said to the Congress, ". . . beauty must not be just a holiday treat, but a part of our daily life."

I return to that theme in this message, which concerns the air we breathe, the water we drink and use, and oceans that surround us, the land on which we live.

These are the elements of beauty. They are the forces that shape the lives of all of us—housewife and farmer, worker and executive, whatever our income and wherever we are. They are the substance of The New Conservation.

Today, the crisis of conservation is no longer quiet. Relentless and insistent, it has surged into a crisis of choice.

Man—who has lived so long in harmony with nature—is now struggling to preserve its bounty.

Man—who developed technology to serve him—is now racing to prevents its wastes from endangering his very existence.

Our environment can sustain our growth and nourish our future. Or it can overwhelm us. History will say that in the 1960s the Nation began to take action so long delayed.

But beginning is not enough. The America of the future will reflect not the wisdom with which we saw the problem, but the determination with which we saw it through.

If we fail now to complete the work so nobly begun, our children will have to pay more than the price of our inaction. They will have to bear the tragedy of our irresponsibility.

The new conservation is work not for some Americans—but for all Americans. All will share in its blessings—and all will suffer if the work is neglected. The work begins with the family. It extends to all civic and community groups. It involves city hall and State capitol. And finally it must engage the concern of the Federal Government.

I urge the Congress to give prompt and favorable consideration to the proposals in this Message.

Source: American Presidency Project. "Papers of Lyndon B. Johnson." [Online information; retrieved 10/1/07.] http://www.presidency.ucsb.edu/ws/index.php?pid=28719.

Waste Management Legislation

The issue of solid waste management that was officially recognized in the 1960s with President Johnson's efforts to control litter alongside the nation's roadways continued with passage of the nation's primary solid waste law, the Resource Conservation and Recovery Act (RCRA) in 1976. The legislation was driven by the momentum of the environmental movement and concerns about the need to recycle more products to conserve natural resources. Solid waste management had traditionally been the purview of local governments, which were considered the responsible agencies for cleaning up this "local" problem. With passage of RCRA, it appeared that the federal government would take a more centralized role in dealing with this perennial environmental problem; the legislation authorized federal grants that would pay for up to 75 percent of demonstration projects. But the law also encouraged newly enhanced disposal techniques such as incineration rather than reducing the amount of waste or recycling, and the Environmental Protection Agency began to shift its emphasis to hazardous, rather than municipal, waste.

As these excerpts from the statute show, Congress was concerned about the combined effects of population and economic growth, the aesthetic and health effects of open dumps and landfills, and risks attributed to uncontrolled pollution. The passage of earlier legislation dealing with air quality and water pollution opened the door for Congress to turn its attention to solid and hazardous waste, as it did in this measure.

The Resource Conservation and Recovery Act of 1976 (excerpts)

The Congress finds with respect to solid waste—

(1) that the continuing technological progress and improvement in methods of manufacture, packaging, and marketing of consumer products has resulted in an ever-mounting increase, and in a change in the characteristics, of the mass material discarded by the purchaser of such products;

(2) that the economic and population growth of our Nation, and the improvements in the standard of living enjoyed by our population,

have required increased industrial production to meet our needs, and have made necessary the demolition of old buildings, the construction of new buildings, and the provision of highways and other avenues of transportation, which, together with related industrial, commercial, and agricultural operations, have resulted in a rising tide of scrap, discarded, and waste materials;

(3) that the continuing concentration of our population in expanding metropolitan and other urban areas has presented these communities with serious financial, management, intergovernmental, and technical problems in the disposal of solid wastes resulting from the industrial, commercial, domestic, and other activities carried on in such areas;

(4) that while the collection and disposal of solid wastes should continue to be primarily the function of State, regional, and local agencies, the problems of waste disposal as set forth above have become a matter national in scope and in concern and necessitate Federal action through financial and technical assistance in the development, demonstration, and application of new and improved methods and processes to reduce the amount of waste and unsalvageable materials and to provide for proper and economical solid waste disposal practices.

Source: U.S. Government Printing Office. [Online information; retrieved 2/13/08.] http://www.access.gpo.gov/uscode/title42/chapter82_.html.

Municipal Solid Waste Recycling

The U.S. Environmental Protection Agency's (EPA) Office of Solid Waste is responsible for coordinating efforts to measure and characterize municipal solid waste (MSW) in a way that provides tools for comparison and improved implementation. The agency publishes a report on U.S. MSW every two years (www.epa.gov/epaoswer), along with various publications that provide guidance for local governments seeking to increase their recycling rate. In order to standardize statistics, the EPA has also developed a list that outlines the scope of what is and what is not MSW, as seen in Table 6.1. The development of standards makes it possible to gauge more closely what types of wastes are being collected as part of the MSW stream and to determine how much recycling is actually taking place.

TABLE 6.1
Scope of Materials Included in the Standard MSW Recycling Rate

Material	What Is MSW?	What Is Not MSW?
Food Scraps	Uneaten food and food preparation; wastes from residences and commercial establishments (restaurants, supermarkets, and produce stands); institutional sources (school cafeterias); and industrial sources (employee lunchrooms)	Food processing waste from agricultural and industrial operations
Glass Containers	Containers; packaging; and glass found in appliances; furniture; and consumer electronics	Glass from transportation equipment (automobiles) and construction and demolition (C&D) debris windows
Lead-Acid Batteries	Batteries from automobiles; trucks; and motorcycles	Batteries from aircraft; military vehicles; boats; and heavy-duty trucks and tractors
Tin/Steel Cans and Other Ferrous Metals	Tin-coated steel cans; strapping; and ferrous metals from appliances (refrigerators); consumer electronics; equipments; and furniture	Ferrous metals from C&D debris and transportation
Aluminum Cans and Other Nonferrous Metals	Aluminum cans; nonferrous metals from appliances; furniture; and consumer electronics; and other aluminum items C&D (foil and lids from bimetal cans)	Nonferrous metals from industrial applications and debris (aluminum siding; wiring; and piping)
Paper	Old corrugated containers; old magazines; old newspapers; office papers; telephone directories; and other paper products, including books, third-class mail, commercial printing, paper towels, and paper plates and cups	Paper manufacturing waste (mill broke) and converting scrap not recovered for recycling
Plastic	Containers; packaging; bags and wraps; and plastics found in appliances; furniture; and sporting and recreational equipment	Plastics from transportation and equipment
Textiles	Fiber from apparel; furniture; linens (sheets and towels); carpets and rugs; and footwear	Textile waste generated during manufacturing process (mill scrap) and C&D projects
Tires	Tires from automobiles and trucks	Tires from motorcycles; buses; and heavy farm and construction equipment
Wood	Pallets; crates; barrels; and wood found in furniture and consumer electronics	Wood from C&D debris (lumber and tree stumps) and industrial process waste (shavings and sawdust)
Yard Trimmings	Grass; leaves; brush and branches; and tree stumps	Yard trimmings in C&D debris
Other	Household hazardous waste (HHW); oil; filters; fluorescent tubes; mattresses; and consumer electronics	Abatement debris; agricultural waste; combustion ash; C&D debris; industrial process waste; medical waste; mining waste; municipal sewage and industrial sludges; natural disaster debris; used motor oil; oil and gas waste; and pre-consumer waste

Source: U.S. Environmental Protection Agency. Office of Solid Waste. 2007. "Scope of Materials Included in the Standard New Recycling Rate." [Online article or information; retrieved October 6, 2007.] www.epa.gov/epaoswer/.

Creating a Market for Recycled Products

In the mid-1970s, at the height of the environmental movement's efforts to rethink waste management strategies, recycling was poised to become an entrenched part of American life. There were expectations that recycled products would become commonplace both in the home and the business sector as technology revealed more ways of reusing products from paper and plastics to glass and tires. One of the reasons why recycling did not fulfill its early promise was the lack of a market for recycled goods. Consumers found that common household recycled content products, such as toilet paper and paper towels, were much more expensive and not as absorbent or soft as the products they were used to using. Paper products, among the most widely available recycled-content items, were often found only in specialty natural foods stores rather than markets and grocery stores. The niche market did not seem especially attractive to producers of these goods.

The federal government responded in 1995, when the EPA issued the first Comprehensive Procurement Guide (CPG) as part of the effort to promote materials made from recycled solid waste. RCRA requires the agency to consider several criteria in determining which items will be listed, and the list applies to procuring agencies (local, state, and federal agencies and their contractors that use federal funds) that spend more than $10,000 a year on that item. In addition, Executive Order 13101 requires the EPA to issue guidance on buying "environmentally preferable products" that have a lesser or reduced effect on human health and the environment when compared with other products and services with the same purpose. The EPA, however, is not authorized to enforce the legislation and its own guidelines.

The initial CPG included five procurement guidelines and 19 products; the November 1997 list added 12 more items, and in January 2000, it designated 18 more. The next update, in April 2004, added seven more items to the procurement list, and there are now eight categories that include recycled-content recommendations for each item.

EPA Comprehensive Procurement Guidelines Products

Construction Products: building insulation productions; polyester carpet; carpet cushion; cement and concrete containing coal fly ash, ground granulated blast furnace slag, cenospheres, silica fume; consolidated and reprocessed latex paint; floor tiles; flowable fill; laminated paperboard; modular threshold ramps; nonpressure pipe; patio blocks; railroad grade crossing surfaces; roofing materials; shower and restroom dividers and partitions; structural fiberboard.

Landscaping Products: compost made from yard trimmings or food waste; garden and soaker hoses; hydraulic mulch; lawn and garden edging; plastic lumber landscaping timbers and posts.

Nonpaper Office Products: binders, clipboards, file folders, clip portfolios, and presentation folders; office furniture; office recycling containers; office waste receptacles; plastic desktop accessories; plastic envelopes; plastic trash bags; printer ribbons; toner cartridges.

Paper and Paper Products: commercial/industrial/sanitary tissue products; miscellaneous papers; newsprint; paperboard and packaging products; printing and writing papers.

Park and Recreation Products: park benches and picnic tables; plastic fencing; playground equipment; playground surfaces; running tracks.

Transportation Products: channelizers; delineators; flexible delineators; parking stops; traffic barricades; traffic cones.

Vehicular Products: engine coolants; rebuilt vehicular parts; re-refined lubricating oils; retread tires.

Miscellaneous Products: awards and plaques; bike racks; blasting grit; industrial drums; manual-grade strapping; mats; pallets; signage; sorbents.

Source: U.S. Environmental Protection Agency. "Comprehensive Procurement Guidelines (CPG)." [Online information; retrieved 10/16/07.] www.epa.gov/epaoswer/non-hw/procure/.

Global Municipal and Hazardous Waste Generation, Collection, and Treatment

Environmental stakeholders are becoming increasingly aware of the magnitude of waste generated on the global level as record keeping and statistical analysis become more technologically advanced. The scale of municipal solid waste generated and collected is especially important because it becomes one of the competing needs for municipal services, considered part of a nation's quality of life. Hazardous waste generation is also a growing problem because of the export and import of materials ranging from toxic byproducts of manufacturing to electronic waste (e-waste) and radioactive materials. The amount of waste a country generates depends on numerous factors, including its gross domestic product, the extent of urbanization, family structures, and lifestyles. The tables below show trends in municipal and hazardous waste generation, collection, and treatment. Although the United Nations compiles statistics on an international basis, only a sampling of results from represented countries with the most complete information is included here in Tables 6.2, 6.3, and 6.4. The indicators are for the latest year available, generally between 2002 and 2005. Caution should be used in comparing countries, as these statistics represent waste collected by or on behalf of municipalities. Waste collected by the informal sector or waste generated in areas not covered by the municipal waste collection system is not included. Hazardous waste statistics include some breaks in time series because of new legislative initiatives and may include figures for biomedical waste and contaminated soil in some countries' inventories, but generally figures are reported according to Basel Convention categories. Wherever possible, like data have been used to make comparisons more reliable.

The Basel Convention

Globalization has come with increased benefits for many nations, and disastrous consequences for others. Among the numerous problems affecting poorer nations, especially those in the Southern Hemisphere, hazardous waste trading is perhaps one of the most insidious. Illegal shipping and transport operations frequently target poor countries in Africa and the Caribbean for

TABLE 6.2
Municipal Waste Collection, by Selected Country

Country	MSW Collected (1000 tons)	Population Served (Percentage)	MSW Collected Per Capita Served (kg/person)
Algeria	8,500	80.0	334
Austria	4,588	100.0	562
Belarus	2,661	85.0	319
Belize	86	51.2	655
Brazil	57,563	76.0	441
Canada	13,375	99.0	423
Cuba	4,416	75.6	519
Denmark	3,618	100.0	675
France	33,963	100.0	561
Greece	4,710	100.0	429
Guatemala	604	30.5	165
Iceland	147	100.0	503
Ireland	2,847	76.0	903
Japan	54,367	99.8	427
Martinique	340	100.0	863
Monaco	40	100.0	1,180
Peru	4,740	75.0	240
Romania	6,865	90.0	341
Singapore	5,088	100.0	1,176
Tunisia	1,316	65.0	203
UK	35,077	100.0	588
US	222,863	100.0	747

Source: United Nations. Department of Economic and Social Affairs, Statistics Division [Online information; retrieved February 15, 2008] http://unstats.un.org/unsd/ENVIRONMENT/municipalwaste.htm.

dumping extremely hazardous wastes of foreign origin, as explained in Chapter 3. Since the 1980s, attempts have been made to create and secure ratification of international agreements to monitor the waste trade, usually because of pressures from international nongovernmental organizations, such as Greenpeace.

The result has been some success, but only in piecemeal fashion. Developed countries usually have some sort of regulatory schemes in place, but the primary regime is the Basel Convention, initially signed in 1989 and entered into force in 1992. As the Preamble to the Convention below illustrates, the agreement attempts to find a middle ground between state sovereignty and local capacity to deal with this growing problem.

TABLE 6.3
Municipal Waste Treatment, by Selected Country

Country	Collected (1000 tons)	Landfilled	Incinerated	Recycled	Composted
Algeria	8,500	99.9%	0.0%	0.1%	0.0%
Austria	4,588	6.7	21.1	26.5	44.7
Belarus	2,661	100.0	0.0	0.0	0.0
Belize	86	100.0	0.0	0.0	0.0
Brazil	57,563	62.7	0.3	1.4	4.1
Canada	13,375	73.3	0.0	26.8	12.5
Cuba	4,416	84.1	0.0	4.8	11.1
Denmark	3,618	5.1	54.0	25.6	15.3
France	33,963	36.0	33.8	15.8	14.3
Greece	4,710	91.9	0.0	8.1	0.0
Iceland	147	72.1	8.8	15.6	8.8
Ireland	2,847	66.1	0.0	33.9	0.0
Japan	54,367	3.4	74.0	16.8	0.0
Martinique	340	67.4	31.9	0.0	0.0
Monaco	40	56.5	n/a	3.7	n/a
Peru	4,740	65.7	n/a	14.7	n/a
Romania	6,865	97.5	0.0	2.5	0.0
Singapore	5,088	15.8	44.8	39.4	0.0
Tunisia	1,316	99.9	0.0	0.0	0.1
UK	35,077	64.3	8.4	17.4	9.3
US	222,863	54.3	13.6	23.8	8.4

Source: United Nations. Department of Economic and Social Affairs, Statistics Division [Online information; retrieved February 15, 2008] http://unstats.un.org/unsd/ENVIRONMENT/wastetreatment.htm.

Preamble to the Basel Convention on the Control of Transboundary Movements of Hazardous Wastes and Their Disposal (March 1989)

The Parties to this Convention,

Aware of the risk of damage to human health and the environment caused by hazardous wastes and other wastes and the transboundary movement thereof,

Mindful of the growing threat to human health and the environment posed by the increased generation and complexity, and transboundary movement of hazardous wastes and other wastes,

Convinced that States should take necessary measures to ensure that the management of hazardous wastes and other wastes including

TABLE 6.4
Hazardous Waste Generation, by Selected Country
(1990–2005, in 1,000 tons)

Country	1990	1995	2000	2001	2002	2003	2004	2005
Armenia	—	—	2.0	1.6	1.2	420.4	544.7	—
Austria	316.8	595.0	1,034.8	1,025.7	920.2	—	1,014.0	—
Belarus	—	90.3	73.0	99.1	116.9	118.5	154.2	—
China	—	—	8,300.0	9,520.0	10,010.0	11,700.0	9,950.0	11,620.0
Denmark	—	179.0	183.4	200.1	247.5	328.3	342.0	340.5
Greece	450.0	350.0	391.0	326.4	352.7	353.8	—	—
Iceland	—	6.0	7.0	8.0	8.0	8.0	8.0	—
Ireland	66.0	248.0	—	491.7	—	—	673.6	—
Italy	3,246.0	2,708.0	3,911.0	4,279.2	5,024.5	5,439.7	5,365.4	—
Monaco	—	0.3	0.3	0.3	0.6	—	—	—
Norway	200.0	650.0	673.0	655.0	—	825.0	940.0	939.0
Singapore	2.3	23.8	29.7	38.4	42.4	43.0	38.2	37.1
Spain	1,708.0	3,394.0	3,063.4	3,222.9	3,222.9	3,222.9	3,534.3	—
UK	2,936.0	2,160.0	5,419.0	5,526.4	5,370.0	4,991.0	5,285.5	—
US	277,339.0	194,225.0	—	37,033.2	—	27,375.0	—	34,788.4

Source: United Nations. Department of Economic and Social Affairs, Statistics Division [Online information; retrieved February 15, 2008] http://unstats.un.org/unsd/ENVIRONMENT/hazardous.htm.

their transboundary movement and disposal is consistent with the protection of human health and the environment whatever the place of disposal,

Noting that States should ensure that the generator should carry out duties with regards to the transport and disposal of hazardous wastes and other wastes in a manner that is consistent with the protection of the environment, whatever the place of disposal,

Fully Recognizing that any State has the sovereign right to ban the entry or disposal of foreign hazardous wastes and other wastes in its territory,

Recognizing also the increasing desire for the prohibition of transboundary movements of hazardous wastes and their disposal in other States, especially developing countries,

Convinced that hazardous wastes and other wastes should, as far as compatible with environmentally sound and efficient management, be disposed of in the State where they were generated,

Aware also that transboundary movements of such wastes from the State of their generation to any other State should be permitted only when conducted under conditions which do not endanger human health and the environment, and under conditions in conformity with the provisions of this Convention,

Considering that enhanced control of transboundary movement of hazardous wastes and other wastes will act as an incentive for their environmentally sound management and for the reduction of the volume of such transboundary movement,

Convinced that States should take measures for the proper exchange of information on and control of the transboundary movement of hazardous wastes and other wastes from and to those States,

Noting that a number of international and regional agreements have addressed the issue of the protection and preservation of the environment with regard to the transit of dangerous goods,

Taking into account the Declaration of the United Nations Conference on the Human Environment (Stockholm, 1972), the Cairo Guidelines and Principles for the Environmentally Sound Management of Hazardous Wastes adopted by the Governing Council of the United Nations Environment Programme (UNEP) by decision 14/30 of 17 June 1987, the Recommendations of the United Nations Committee of Experts on the Transport of Dangerous Goods (formulated in 1957 and updated biennially), relevant recommendations, declarations, instruments and regulations adopted with the United Nations system and the work and studies done within other international and regional organizations,

Mindful of the spirit, principles, aims and functions of the World Charter for Nature adopted by the General Assembly of the United Nations at its thirty-seventh session (1982) as the rule of ethics in respect of the protection of the human environment and the conservation of natural resources,

Affirming that States are responsible for the fulfilment of their international obligations concerning the protection of human health and protection and preservation of the environment, and are liable in accordance with international law,

Recognizing that in the case of a material breach of the provisions of this Convention or any protocol thereto the relevant international law of treaties shall apply,

Aware of the need to continue the development and implementation of environmentally sound low-waste technologies, recycling options, good house-keeping and management systems with a view to reducing to a minimum the generation of hazardous wastes and other wastes,

Aware also of the growing international concern about the need for stringent control of transboundary movement of hazardous wastes and other wastes, and of the need as far as possible to reduce such movement to a minimum,

Concerned about the problem of illegal transboundary traffic in hazardous wastes and other wastes,

Taking into account also the limited capabilities of the developing countries to manage hazardous wastes and other wastes,

Recognizing the need to promote the transfer of technology for the sound management of hazardous wastes and other wastes produced locally, particularly to the developing countries in accordance with the spirit of the Cairo Guidelines and decision 14/16 of the Governing Council of UNEP on Promotion of the transfer of environmental protection technology,

Recognizing also that hazardous wastes and other wastes should be transported in accordance with relevant international conventions and recommendations,

Convinced also that the transboundary movement of hazardous wastes and other wastes should be permitted only when the transport and ultimate disposal of such wastes is environmentally sound.

Source: Secretariat of the Basel Convention. [Online information; retrieved 2/16/08.] www.basel.int.

Hazardous Waste Dumping

Late at night on August 19, 2006, the vessel *Probo Koala* illegally dumped hazardous waste at various sites in Abidjan, Cote d'Ivoire, including areas close to water sources and a lagoon, killing several persons and sickening thousands of residents. An investigation by the government's antipollution laboratory later found that the dumped material included volatile pollutants that reach humans and animals through the air and drinking water and accumulate in the food chain. Several reports indicated that the dumping had been authorized by public authorities who believed it was sewage, resulting in a political crisis that led to the dissolution of the African nation's cabinet. Officials announced that they did not have the capacity to deal with the damage, requesting international assistance estimated to reach more than $13.5 million for the short- and medium-term response to the dumping.

Three months later, the United Nations under-secretary general, who also serves as the director of the United Nations Environment Programme, spoke at the opening of the Eighth Conference of the Parties to the Basel Convention about the incident. The international agreement, outlined in Chapter 3, is considered to be an initial effort to curb hazardous waste dumping, especially among poor countries that are not equipped to deal with this problem. In this excerpt from the opening address, the incident is raised in the context of both the convention and the growing problem of electronic waste.

Speech by Achim Steiner to the Opening of the Eighth Conference of the Parties to the Basel Convention, Nairobi, Kenya (November 27, 2006)

This week's meeting will address many issues not least the rising tide of electronic or e-waste, but it would be impossible for me not to refer to the tragic events in Cote D'Ivoire.

This case of hazardous waste dumping in one of the poorest countries on the globe serves as a reminder of the importance of the Basel Convention and the need to re-invigorate and re-new its vital regional and global role.

It also serves as a reminder that even the best laws are only as strong as the enforcement mechanisms and willingness of governments to act.

Ladies and gentlemen, the need for Basel is ever more evident in this globalized world and not just because of Cote D'Ivoire.

Accelerating trade in goods and materials across borders and Continents is one of the defining features of the early 21st century.

Another is the globalized phenomenon of consumerism and what one might call "built in obsolescence"—the relative cheapness of high technology products like mobile phones and computers—the way fashion is driving the purchasing and discarding of products in a way unknown a generation ago.

Consumerism is driving economies but also drives a growing mountain of e-waste not only from North to the South but South-South—waste with a wide range of pollutants from heavy metals to chlorine compounds. . . .

One of the great challenges of our time is to collectively agree on what is waste and what are second-hand products—this question extends to end-of-life ships as much as to electronic goods.

Unless we get to grips with this, we are always going to be like the proverbial dog chasing its tail. . . .

I sincerely hope that the tragedy in Cote D'Ivoire and the challenges of e-waste will serve as a wake up call to the Parties of the Basel Convention and other related treaties.

The failings in the Cote D'Ivoire underscores the need for the international environmental and development community to echo to the High Level Panel's conclusions by focusing and intelligently deploying the legal and other instruments at our disposal. . . .

We have this instrument called Basel. Let us support it practically, politically and financially to achieve its full potential in our common pursuit of overcoming poverty and achieving sustainability on this wonderful planet Earth.

Source: United Nations Environment Programme. [Online article; retrieved 10/16/07.] http://www.unep.org/Documents.Multilingual/Default.print.asp?DocumentID=495&ArticleID=5433&1=en.

National Priorities List of Waste Site Actions

The U.S. Environmental Protection Agency is responsible for developing a list of hazardous waste sites under the Superfund program established by the Comprehensive Environmental Response, Compensation and Liability Act in 1980. The EPA conducts a review under the Hazard Ranking System and addresses public comments to determine whether the site warrants further investigation to assess the nature and extent of human health and environmental risks prior to any remedial actions. The National Priorities List (NPL) includes three types of locations: sites proposed for listing, sites finalized on the list, and sites deleted from the NPL. The data in Table 6.5, covering the fiscal years 1995 through 2007, show that far more sites have been proposed and then finalized on the NPL than have been deleted through completion of Superfund remediation projects.

"Seven New (Garbage) Wonders of the World"

Many lists identify what are considered the "wonders of the world," both ancient and modern. Most represent architectural achievements such as the pyramids of Giza, the only structure still standing from the original list of seven wonders, or in contemporary lists, sites such as the Sydney Opera House or the Statue of Liberty in New York. Writer Aaron Labaree has developed his own list of the seven wonders of the garbage world in conjunction with Ann Leonard, an anti-incinerator advocate who has traveled widely over the last 20 years visiting trash and waste sites. The following list is extracted from Labaree's 2006 article on the world's waste.

TABLE 6.5
National Priorities List Site Actions, by Year (1995–2007)

Year	Proposed	Finalized	Deleted	Partial Deletions*	Construction Completions
1995	9	31	25	0	68
1996	27	13	34	0	64
1997	20	18	32	6	88
1998	34	17	20	7	87
1999	37	43	23	3	85
2000	40	39	19	5	87
2001	45	29	30	4	47
2002	9	19	17	7	42
2003	24	20	9	7	40
2004	26	11	16	7	40
2005	12	18	18	5	40
2006	10	11	7	3	40
2007	5	5	6	3	11

*These statistics represent the total number of partial deletions by fiscal year and may include multiple partial deletions at a site.

Source: U.S. Environmental Protection Agency [Online information; retrieved September 11, 2007.] http://www.epa.gov/superfund/sites/query/queryhtm/nplfy.htm.

1. The Eastern Garbage Patch, Pacific Ocean

Turning and turning in a widening gyre is a field of floating plastic trash bigger than Texas. Probably the world's largest dump, the EGP is about halfway between San Francisco and Hawaii, an area where weak winds and sluggish currents cause flotsam and jetsam from around the Ocean to collect. The trash comes mostly from land but also from loads lost or jettisoned by ships. One researcher recounts sailing for a week and being able to see plastic trash floating from horizon to horizon.

2. Fresh Kills Landfill, Staten Island, USA

As every schoolboy knows, there are two man-made structures visible from space: one kept the Mongol hordes out of China, one received New York's garbage for fifty years. Fresh Kills is now sealed up and the City of New York plans to turn it into a park, but its sheer size earns it a place on this list. It is one of the largest man-made structures on the planet, 225 feet at its highest point and covering 2,200 acres; during its peak intake in the mid-eighties, it received 29,000 tons of garbage a day.

3. Roro Asbestos Dump, Jkarkhand, India

Asbestos mines in the Roro hills in central India closed down in 1983, and since then a gigantic pile of mining waste has lain there unattended, slowing spreading downhill into nearby paddy fields. About 700,000 tons of leftover asbestos and the rock it was mined from form a toxic mini-mountain on the hills. While Roro can't compete with the other Wonders for scale, the striking image it presents of an open geologic-formation-size heap of a substance which in this country is handled by people in full-body suits puts it on the list. Preliminary studies show major health effects on the local population, although the results are complicated by the fact that many of them used to work in the mine itself.

4. Smoky Mountain, Manila, Philippines

Officially shut down in 1995, this garbage dump (its picturesque name comes from the clouds of methane that rose above it when it was active) for decades provided a livelihood for thousands of people who lived on or near it. Smoky Mountain makes the list because of its fame; it has probably been surpassed in size and relevance by other sites in Manila, where as many as 150,000 people, many of them migrants from the provinces, make their living scavenging trash. Men, women, and children alike live by the dumps in some of the world's worst slums. In the year 2000, over three hundred people were killed when one of the faces of Payatas dump collapsed.

5. Shipbreaking Yards, Alang, India

Every year hundreds of big ships are retired and sent to Asia for scrap. They are often dismantled on what were once pristine beaches, by workers with no protection from toxic materials. Conditions at Alang have somewhat improved in the last few years and it may soon lose its place on this list to an enormous planned shipbreaking site on Vodarevu Beach, which would cover an area of over 160 football fields. But it's still impressive. Leonard says of her visit a few years ago, "It is the closest thing I have ever seen to what I picture hell to look like. Miles of filthy shoreline, gigantic tankers beached and being stripped of any valuable parts. There is smoking oil, huge chunks of metals, flames all over. Thousands of barely clad workers crawl all over these hulking ships with flame torches or primitive tools. Big pieces of metal fall and many [workers] lose their limbs. Some of these sites average a worker death a day."

6. Yucca Mountain, Nevada, USA

Exploratory work has begun on this repository for nearly all of the U.S.'s nuclear waste, which if completed would be a wonder that no living thing would ever see. Over 70,000 tons of radioactive material must be isolated for at least 10,000 years to avoid a potential toxic cataclysm. To do this, canisters of waste would be carried into the mountain by rail car, and remote-controlled equipment would put the canisters on supports in the underground tunnel. . . . 1,000 feet below the mountain's surface. The project is set to be completed by 2017, and costs are estimated in the tens of billions. Nevadans are not happy about the idea and have not been won over by the cartoon character used to promote the plan to children, Yucca Mountain Johnny.

7. Electronic Waste Dumps, Guiyu, China

This small area near China's East Coast is a major destination for North America's electronic waste. While not as visually impressive as some of the other Wonders, the villages in the area process an incredible amount of computer waste. As many as 100,000 people are employed in this way, most of them rural migrant workers. In every neighborhood, in back streets, outside the town by the river, piles of electronic waste are dismantled by hand. In one small village, the residents make their living entirely by burning wires to recover copper. Such huge quantities of toxic e-waste have passed through the area that sediment samples taken near the river show levels of dangerous metals like tin and lead at hundreds of times the EPA threshold.

Source: Labaree, Aaron. 2006. "Seven New (Garbage) Wonders of the World." *New York Inquirer.* [Online article; retrieved 4/11/07.] http://www.nyinquirer.com/nyinquirer/2006/11/seven_new_garba.html.

World's Most Polluted Places, 2006

In 2006, the Blacksmith Institute issued the first-ever list of the world's most polluted places, identifying the top 10 locations where toxic waste severely affects human health, especially that of children. The list was compiled from more than 300 suggestions that had been submitted to the group, which were then reduced to 35 sites for further research. The environmental and health experts who serve on the organization's technical advisory board used a scoring system developed for this purpose to weigh pollution conditions in terms of their impact on health, including the size of the affected population. With the exception of Chernobyl, Ukraine, most Americans are not likely familiar with any of these waste sites, which represent the "Brown Agenda" of long-term pollution.

The final list, in alphabetical order, includes the following locations and a brief explanation of why they were selected and what is being done.

Chernobyl, Ukraine

The site of the world's worst nuclear disaster on April 26, 1986, was selected because of its residual environmental impact, as well as its potential to further affect such an extensive region and population. An estimated 5.5 million people were initially affected when testing at the nuclear power plant triggered a meltdown of the reactor's core. Currently, a 19-mile, uninhabited exclusion zone surrounds the facility, where the reactor is buried in a concrete casing designed to absorb radiation and contain the remaining nuclear fuel. Implementation of an integrated radioactive waste management program to ensure consistent management and facility capacity needs to be assessed before further development can take place. Remediation costs have been estimated at hundreds of billions of dollars.

Dzerzhinsk, Russia

Cold War–era chemicals and toxic by-products used in chemical weapons manufacturing have led to a reduction in life expectancy in this area, where an estimated 300,000 people may be affected. The average life expectancy for men is 42 years and for women is 47 years. Almost 300,000 tons of chemical waste were improperly disposed of from 1930 to 1998, leading to groundwater contamination. The chemicals have turned the water into a white sludge containing dioxins and high levels of phenol that are reportedly 17 million times the safe limit. Efforts are being undertaken to install new water systems and to evaluate the extent of groundwater contamination; the city's residents are still employed in factories that produce toxic chemicals.

Haina, Dominican Republic

Also known as Bajos de Haina, this closed-down automobile battery recycling center is a waste hot spot where lead poisoning potentially affects 85,000 people. Birth deformities; eye damage; learning and personality disorders; and, in some cases, death from lead poisoning have been reported at a higher than normal rate because of past contamination from the Metaloxa company's operations, which ended in March 1997. Cleanup activity is in the early planning stages; the nation's Secretary of Environment and Natural Resources post was not created until 2000.

Kabwe, Zambia

Nearly a quarter of a million people are potentially affected by the lead and cadmium waste and dust in Kabwe, the second-largest city in Zambia. It is situated in the country's copper belt, where rich deposits of lead were discovered in 1902 and smelters were built to process the ore. Until 1994, the operations were not regulated by the government, during which time heavy metals were released into the air as dust particles and slag heaps of mining wastes were created. Children still play in the contaminated soil, and young men scavenge the mines for scraps of metal. The levels of lead found in residents' blood are as much as 20 times higher than permissible blood levels in the United States. The cleanup of the area is still in the primary stages and must begin with educating the community about the risks of lead poisoning. Residents in some parts of the area may need to relocate; the World Bank has approved a $20 million grant to clean up the city.

La Oroya, Peru

The Missouri-based Doe Run Corporation owns a poly-metallic smelter in this town in the Peruvian Andes, home to about 35,000 people. The plant began operating in 1922, and since that time, residents have been exposed to toxic levels of lead. An estimated 99 percent of the children in La Oroya have blood lead levels that exceed acceptable amounts, the government reports, a condition that can cause diminished mental development. Sulfur dioxide concentrations also exceed the World Health Organization's emissions standards by 10-fold. Vegetation in the area has been destroyed by acid rain created by these emissions, and lead and other metals are still being deposited while the plant is in operation. The extent of soil contamination has not been studied, and no plan for reduction of emissions has been agreed upon or implemented.

Linfen, China

Linfen, located in Shanxi Province in the heart of China's coal belt, provides about two-thirds of the nation's energy. China's urgent need for coal has resulted in hundreds of illegal and unregulated mining operations. The waste and emissions from the coal mines, steel factories, and tar factories produce choking dust that has fouled waterways; water is rationed to the extent that even the provincial capital receives water for only a few hours each day. Linfen is considered the city with the worst air quality in China, with local clinics seeing a growing number of cases of bronchitis, pneumonia, and lung cancer. Arsenicosis, an environmental disease caused by drinking water with elevated concentrations of arsenic that results in skin lesions, peripheral vascular disease, hypertension, blackfoot disease, and a high risk of cancers, is also being seen in epidemic proportions. No information is available indicating whether this site is being cleaned up at this time.

Mailuu-Suu, Kyrgyzstan

Scattered throughout this area are 23 tailing dumps and 13 waste rock dumps, where a uranium plant was in operation from 1946 to 1968. More than 10,000 tons of uranium ore were produced and processed to build the former Soviet Union's first atomic bomb. An estimated 1.96 million cubic meters of radioactive mining waste threatens one of Central Asia's most fertile and densely populated valleys. Compounding the waste problem is the fact that this is an area of high seismic activity, and natural hazards such as landslides and mudflows could potentially exacerbate the failure of waste containment. It is feared that an earthquake or other event could disturb one of the dumps to expose radioactive material; in 2002, a huge mudslide blocked the course of the Mailuu-Suu River near the mine tailings. A similar event could contaminate water sources, affecting hundreds of thousands of people. The World Bank estimates the cost of isolating and protecting the uranium waste at $11.76 million.

Norilsk, Russia

In 1935, Norilsk was founded as a slave labor camp. It is the northernmost city in Russia, with a population of about 135,000, and the second-largest city above the Arctic Circle. It is home to the world's largest heavy metal smelting complex and the world's largest producer of nickel. Because of its location, which is off-limits to foreigners, it has been difficult for researchers to know much about the facilities there, although the site is considered one of the most polluted in Russia. It has been shut down to outside visitors since November 2001, one of 90 "closed towns" in Russia where Cold War–era secrecy persists.

Ranipet, India

Ranipet is a medium-size town in an area where 3.5 million people are at risk from a facility that manufactures sodium chromate, chromium salts, and chromium sulfate powder used in local tanning operations. The Tamil Nadu Pollution Control Board estimates that about 1.5 million tons of solid wastes accumulated over two decades of plant operations, stacked in an open yard, 3 to 5 meters high on 2 hectares of land. The facility was closed down by the government in 1996 after farmers complained the fields were toxic to agriculture and that they became sickened when they came into contact with the water they used to irrigate their fields. Studies show the solid waste could be dealt with by encapsulating it to prevent further leaching into the soil and groundwater, and plans are underway to design a remediation program for the site.

Rudnaya Pristan/Dalnegorsk, Russia

These two towns are located in the Russian Far East and are home to an old lead smelter and mining activities that have produced dangerous levels of lead in the soil, dust, air, and drinking water. Since 1930, no attempts have been made to address the 90,000 residents' health concerns, and most are unaware of the risks they face. Some use old casings from submarine batteries that were recycled by the smelter in order to collect rain to water their gardens. The site has now been voluntarily shut down, and children in the towns have had their blood tested to look for excessive lead concentrations. Funding has also been obtained to support a public education program.

Source: Blacksmith Institute. 2006. "World's Worst Polluted Places." [Online information; retrieved 6/21/07.] www.blacksmithinstitute.org.

TABLE 6.6
Supply and Recovery of Paper and Paperboard, 1993–2006
(in 1,000 tons)

Year	Supply	Recovered	Recovery Rate
1993	91,538	35,460	38.70%
1994	95,718	39,691	41.50
1995	95,971	42,189	44.00
1996	94,529	43,076	45.60
1997	99,557	43,989	44.20
1998	101,183	45,076	44.60
1999	105,316	46,818	44.50
2000	102,810	47,311	46.00
2001	97,395	46,996	48.30
2002	98,949	47,645	48.20
2003	98,016	49,255	50.30
2004	101,882	50,287	49.40
2005	99,618	51,272	51.50
2006	100,198	53,488	53.40

Source: American Forest and Paper Association [Online article; retrieved September 11, 2007.] http://stats.paper-recycles.org.

How Much Paper Is Recycled? What Happens to Old Newspapers?

Although the statistics on recycling of many materials shows a general decline, paper and paperboard recovery reached record levels in 2006, increasing to more than 53 percent. The U.S. figures are nearly 8 percent higher than a decade before, when levels were less than 50 percent. Old newspapers can have a second life, too, according to the American Forest & Paper Association (http://stats.paperrecycles.org), which studies recycling patterns. They can be used to produce a variety of new products, as shown in Table 6.6 and Table 6.7. About one-fourth of the nation's old newspaper is exported, while one-third is made back into newsprint.

What Do You Do with Waste Carpet?

Chapter 2 explores the problems of dealing with a unique type of recyclable—waste carpet—and the efforts that began in 2002 to increase the amount of carpet that was recycled or diverted from

TABLE 6.7
Use Statistics for Old Newspapers, 2006
(in 1,000 tons)

Product	Consumption by Share of End Use	Share of Total
Containerboard	1	0.00%
Recycled paperboard	1,172	12.30%
Tissue	698	7.40%
Newsprint	3,213	33.60%
Printing-Writing	353	3.70%
All Other	1,615	16.90%
Net Exports	2,513	26.30%
TOTAL	9,565	100.00%

Source: American Forest & Paper Association [Online information; retrieved September 11, 2007.] http://stats.paperrecycles.org/index.php?graph=oldnews&x=70&y=11.

TABLE 6.8
Postconsumer Carpet Recycling and Diversion, 2002–2006

	Millions of Pounds					Percent of Total Discards				
	2002	2003	2004	2005	2006	2002	2003	2004	2005	2006
Total Discards	4,678	4,828	4,537	5,038	5,261	—	—	—	—	—
Recycled	46.2	86.8	98.4	194.3	239.5	1.0%	1.8%	2.2%	3.9%	4.6%
Diverted	57.2	93.7	108.2	224.6	260.9	1.2%	1.9%	2.4%	4.5%	5.0%

Source: Carpet America *Annual Report,* May 17, 2007 [Online information; retrieved July 10, 2007.] http://www.carpetrecovery.org/pdf/annual_report/06_CARE-annual-rpt.pdf.

landfills by 2012. The industry admits that the first five years of the project, the Carpet America Recovery Effort, have not met its expectations, but the agreement that was reached by the parties does indicate the potential for stewardship of this bulky and expensive-to-handle waste product. Table 6.8 indicates the project's goals and successes thus far.

Recycling Fun Facts

The Can Manufacturers Institute (CMI) is a national industry trade group that represents the metal and composite can manufacturing sectors of business. The group was formed in 1938 to serve as the voice of its members in influencing legislative, regulatory, and administrative policy, including environmental awareness. CMI also promotes community recycling involvement and provides statistical information on can production. This list of 15 interesting tidbits about cans and recycling provides the context for understanding the role of aluminum and metal cans in the waste stream.

1. Recycling aluminum cans saves 95 percent of the energy used to make aluminum cans from virgin ore.
2. Aluminum cans are the most recycled beverage container in the world, with 105,800 cans recycled every minute in the United States.
3. Used aluminum cans are recycled and returned to a store shelf as a new can in as few as 60 days.
4. Americans earn about $1 billion a year recycling aluminum cans.
5. The aluminum can recycling rate is better than one out of two cans.
6. Using recycled aluminum beverage cans to produce new cans allows the aluminum can industry to make up to 20 times more cans for the same amount of energy.
7. Since 1972, an estimated 1,099 billion aluminum cans have been recycled, which, if placed end to end, could stretch to the moon and back about 174 times.
8. The average employee consumes 2.5 cans of soda each day at work.
9. Recycling one aluminum can saves enough energy to run a television for three hours.
10. A baby is born every three seconds, and in that same time, 140 cans are produced.
11. The average American family recycles 150 six-packs of aluminum cans a year.
12. Americans drink an average of 380 beverages in aluminum cans each year.
13. Recycling 1 ton of aluminum cans saves the equivalent in energy of 2,350 gallons of gasoline.
14. Recycling 1 ton of aluminum cans saves the equivalent of the amount of electricity used by the typical home over a period of 10 years.

15. Recycling aluminum creates 97 percent less water pollution than producing new metal from ore.

Source: Can Manufacturers Institute. "Recycling Fun Facts." [Online information; retrieved 9/11/07.] http://www.cancentral.com/funFacts.cfm.

U.S. Beverage Container Recycling Scorecard

Containers and packaging form the largest segment of municipal solid waste and beverage containers comprise nearly 15 percent of all packaging. The nonprofit Container Recycling Institute (CRI) and the social change organization As You Sow produced a report, *Waste and Opportunity, U.S. Beverage Container Recycling Scorecard and Report,* that analyzes the recycling loop from the production, sale, and recovery of beverage containers. The report notes that increased recycling efforts have been unable to match the increases in beverage sales, resulting in a decline in recycling from 53.5 percent in 1992 to 33.5 percent in 2004. These numbers are important because organizations are trying to promote corporate accountability, social justice, and environmental protection. Groups such as the CRI are also attempting to shift the costs associated with manufacturing, recycling, and disposing of container and packaging waste (both social and environmental) from government and taxpayers to producers and consumers.

The organizations' October 2006 study, authored primarily by Nishkita Bakashi, the research director of As You Sow's Corporate Social Responsibility Program, graded efforts by beverage producers on four core criteria: the inclusion of recycled content in beverage containers; involvement in beverage container recovery; involvement in container recycling; and source reduction of plastics, aluminum, and glass. Twelve beverage companies were graded using publicly available information from annual reports and Web sites and on the basis of a survey used in the study, as seen in Table 6.9. PepsiCo led the companies because it met its goal of using 10 percent recycled content in its plastic carbonated soft drink and water bottles in the United States by the end of 2005. But overall, virtually no action is being taken to significantly increase beverage container recovery and reduce waste.

TABLE 6.9
U.S. Beverage Container Recycling Scorecard and Report, 2006

Company	Recycled Content	Recovery and Recycling	Source Grade Reduction	GPA
PepsiCo	B	C+	D+	2.3
Coca-Cola	D	C+	B	2.1
Miller Brewing	D	D–	D	0.9
New Belgium Brewery	C	F	F	0.7
Coors	D+	D–	F	0.7
Anheuser-Busch	F	D	D	0.7
Polar Beverages	F	D	F	0.6
Starbucks	D	F	F	0.3
Nestle Waters	F	F	F	0.1
Cadbury Schweppes	F	F	F	0.0
Cott	F	F	F	0.0
National Beverage	F	F	F	0.0

Source: U.S. Beverage Container Recycling Scorecard and Report, October 2006 [Online information; retrieved September 11, 2007.] http://www.container-recycling.org/assets/pdfs/reports/2006-scorecard.pdf.

The Nuclear Waste Debate

Although nuclear power has a number of advantages in helping the world deal with its demand for energy, controversies over how to deal with nuclear waste have limited its use and development in many nations, including the United States, as explained in Chapter 3. Nuclear waste is stored in a variety of facilities, such as independent spent fuel storage installations in 30 states. The Nuclear Regulatory Commission deals with these sites, and the U.S. Department of Energy manages another group of sites that handle low-level waste from hospitals, universities, and power plants that fall into the lowest-hazard class (waste that decays to nonradioactive levels within 100 years). Three active, low-level-designated waste sites are in operation in the United States: Richland, Washington; Barnwell, South Carolina; and Clive, Utah. These facilities are controversial because they accept low-level waste from other states through multistate compacts. The site in Washington accepts waste from 11 northwestern and Rocky Mountain states; the South Carolina facility has been accepting waste from 38 states and Washington, D.C., but in July 2008 began limiting the waste it accepted to shipments from New Jersey, Connecticut, and in-state

TABLE 6.10
Locations of U.S. Independent Spent Fuel Storage Installations (by State, as of February 2008)

Alabama: Brown's Ferry, Farley	Minnesota: Prairie Island
Arizona: Palo Verde	Mississippi: Grand Gulf
Arkansas: Arkansas Nuclear	Nebraska: Ft. Calhoun
California: Diablo Canyon, Rancho Seco, San Onofre, Humboldt Bay	New Jersey: Hope Creek, Oyster Creek
	New York: James A. FitzPatrick
Colorado: Fort St. Vrain	North Carolina: McGuire
Connecticut: Haddam Neck, Millstone	Ohio: Davis-Besse
Georgia: Hatch	Oregon: Trojan
Idaho: Department of Energy, Idaho Spent Fuel Facility	Pennsylvania: Susquehanna, Peach Bottom
Illinois: GE Morris, Dresden, Quad Cities	South Carolina: Oconee, H. B. Robinson
Iowa: Duane Arnold	Tennessee: Sequoyah
Louisiana: River Bend	Utah: Private Fuel Storage
Maine: Maine Yankee	Virginia: Surry, North Anna
Maryland: Calvert Cliffs	Washington: Columbia Generating Station
Massachusetts: Yankee Rowe	Wisconsin: Point Beach
Michigan: Big Rock Point, Palisades	

Source: U.S. Nuclear Regulatory Commission [Online article or information; retrieved February 15, 2008.] www.nrc.gov/waste/spent-fuel-storage/locations.html.

sources. This means that the privately owned Utah facility will be the only one accepting waste from nuclear generators in 36 states.

Yucca Mountain

One of the milestones in the development of U.S. nuclear waste policy was a decision by President George W. Bush to designate Yucca Mountain, Nevada, as a repository for spent nuclear fuel and waste, as seen in the February 15, 2002, White House message that follows. The action, although not unexpected at the time, came as the result of contentious debate over the Nuclear Waste Policy Act and efforts by Nevada's political leaders to stop the facility from being sited in their state. Gov. Kenny Guinn, who testified frequently before Congress on the Yucca Mountain issue, has been one of the harshest critics of the project. In the selection excerpted from his April 25, 2002, statement before two congressional subcommittees (appearing below the president's message), Guinn makes clear his disagreements with both the science behind the project and the long-term issues it poses. He calls upon Congress to allow his gubernatorial veto of the site selection to stand. Political leaders in Nye County, Nevada, in contrast, have been among the few who have favored the Yucca Mountain site. The July 20, 2004 resolution, of the Nye County Board of County Commissioners (the third excerpt in this section) portrays an entirely different perspective of the repository and the officials' support for the project. The Nuclear Age Peace Foundation in Santa Barbara, California, provides still another view of the Yucca Mountain project and reasons why it opposes the Department of Energy's plan to transport waste to the facility (the final selection).

Excerpts from a Letter from the President to the Speaker of the House of Representatives and the President of the Senate, February 15, 2002

Dear Mr. Speaker:

In accordance with section 114 of the Nuclear Waste Policy Act of 1982, 42 U.S.C. 10134 (the "Act"), the Secretary of Energy has recommended approval of the Yucca Mountain site for the development at that site of a repository for the geologic disposal of spent nuclear fuel and high level nuclear waste from the Nation's defense activities. As is required by the Act, the Secretary has also submitted to me a comprehensive statement of the basis of his recommendation.

Having received the Secretary's recommendation and the comprehensive statement of the basis of it, I consider the Yucca Mountain site qualified for application for a construction authorization for a repository. Therefore I now recommend the Yucca Mountain site for this purpose. In accordance with section 114 of the Act, I am transmitting with this recommendation to the Congress a copy of the comprehensive statement of the basis of the Secretary's recommendation prepared pursuant to the Act. The transmission of this document triggers an expedited process described in the Act. I urge the Congress to undertake any necessary legislative action on this recommendation in an expedited and bipartisan fashion.

Proceeding with the repository program is necessary to protect public safety, health, and the Nation's security because successful completion of this project would isolate in a geologic repository at a remote location highly radioactive materials now scattered throughout the Nation. In addition, the geologic repository would support our national security through disposal of nuclear waste from our defense facilities.

A deep geologic repository, such as Yucca Mountain, is important for our national security and our energy future. Nuclear energy is the second largest source of U.S. electricity generation and must remain a major component of our national energy policy in the years to come. The cost of nuclear power compares favorably with the costs of electricity generation by other sources, and nuclear power has none of the emissions associated with coal and gas power plants.

This recommendation, if it becomes effective, will permit commencement of the next rigorous stage of scientific and technical review of the repository through formal licensing proceedings before the Nuclear Regulatory Commission. Successful completion of this program will also redeem the clear Federal obligation safely to dispose of commercial spent nuclear fuel that the Congress passed in 1982.

This recommendation is the culmination of two decades of intense scientific scrutiny involving application of an array of scientific and technical disciplines necessary and appropriate for this challenging undertaking. It is an undertaking that was mandated twice by the Congress when it legislated the obligations that would be redeemed by successful pursuit of the repository program. Allowing this recommendation to come into effect will enable the beginning of the next phase of intense scrutiny of the project necessary to assure the public health, safety, and security in the area of Yucca Mountain, and also to enhance the safety and security of the Nation as a whole.

Sincerely,
George W. Bush

Source: White House. [Online press release; retrieved 4/17/07.] www.whitehouse.gov/news/releases/2002.

Excerpts from the Statement of Kenny C. Guinn, Governor of the State of Nevada, before the U.S. House of Representatives, Committee on Transportation and Infrastructure Subcommittees on Railroads and Transportation and Hazardous Materials, April 25, 2002

As is well known by this time, Nevada considers the Yucca Mountain project to be the product of extremely bad science, extremely bad law, and extremely bad public policy. Moreover, implementing this ill-conceived project will expose tens of millions of Americans to unnecessary nuclear transport risks. For these reasons we in Nevada believe, and ask, that Congress should take no further action with respect to the Yucca Mountain project.

I would like to briefly call the Committee's attention to a new document, a key document, which has now appeared from within the scientific community that excoriates the scientific work of the Department of Energy (DOE) in connection with Yucca Mountain. Numerous independent scientific reviewers have now evaluated the project during the past year, and all have reached the same conclusion: There is nowhere near enough information to certify the suitability of the Yucca Mountain site for high-level nuclear waste disposal, and the information that is available suggests the site is woefully unsuitable geologically.

This latest report, however, reaches shocking new conclusions. It is a peer review report commissioned by DOE from the International Atomic Energy Agency and the Nuclear Energy Agency (IAEA) of the Organization for Economic Cooperation and Development (OECD). These agencies assembled some of the world's leading scientists to evaluate, over several months, the total system performance of Yucca Mountain as represented by DOE and its computer models. Among other things, these leading scientists concluded that DOE lacks sufficient information even to build a model to predict the suitability and hydrogeologic performance of the proposed repository. According to the peer review group, the water flow system at Yucca Mountain is "not sufficiently understood to propose a conceptual model for a realistic transport scenario."

It is truly amazing to me, as an elected official, that DOE commissioned this peer review report many months ago, and then made a final "site suitability" determination to the President and the Congress in spite of its stunning conclusions. It shows once again, in my view, that politics has long prevailed over science when it comes to Yucca Mountain. This is another reason for Nevada to redouble its efforts to stop this project—government bureaucrats seem unable to pull the plug, even in the face of shocking independent evidence that the science is bad or nonexistent.

Today, the President's recommendation to move forward with Yucca Mountain is heading down the path to finality, and only the Congress can stop it by choosing not to override my recent, Congressionally-authorized, site veto. If the matter of site suitability really were up to the NRC, Nevada and the scores of independent scientists alarmed by DOE's premature and falsely based site recommendation would be considerably reassured. But such is not the case.

If Congress overrides my veto and simply punts to the NRC, the suitability of the Yucca Mountain site will never be independently reviewed by any government authority, barring a court order. We will seek that court order, but we believe Congress should accept its responsibility, recognize that the Yucca Mountain project is fatally flawed on numerous fronts, and not act to override my veto.

Source: Office of Governor Kenny Guinn. [Online press release; retrieved 5/1/02.] http://gov.state.nv.us.

Nye County Resolution No. 2004-25

RESOLUTION CONCERNING THE INTENT OF NYE COUNTY TO TAKE ACTION TO MAXIMIZE THE SAFETY, ECONOMIC OPPORTUNITY AND SUCCESSFUL OUTCOME OF THE YUCCA MOUNTAIN REPOSITORY AND TRANSPORTATION SYSTEM BY ACTIVELY AND CONSTRUCTIVELY ENGAGING ALL RELEVANT PARTIES.

WHEREAS the Nuclear Waste Policy Act of 1982 as amended designates Yucca Mountain, located in Nye County, Nevada, as the only site for consideration as the nation's repository for high-level nuclear waste and spent fuel; and

WHEREAS the site has been determined to be a suitable location for a repository, the U.S. Court of Appeal dismissed all challenges to the site selection of Yucca Mountain, the scientific basis for the selection process and the constitutionality of the resolution approving Yucca Mountain; and

WHEREAS the Department of Energy is preparing a license application for the repository and expects to begin operation beginning in 2010; and

WHEREAS the Department intends to use rail transportation, the mode of transportation Nye County prefers, to the maximum extent possible and the Department has made progress in planning the transportation by selecting the Caliente route; and

WHEREAS the Department is beginning the process of identifying repository and transportation facilities which could be located offsite and is considering other means of maximizing local economic opportunity; and

WHEREAS the Nye County "Community Protection Plan" has established a vision for protecting the community and for the local development of synergistic economic, scientific and education activities for management and possible future reuse of material which will be stored at Yucca Mountain; and

WHEREAS it is just such a vision for the Yucca Mountain Project that offers the best long term prospect for converting long-standing resistance and mistrust within the state of Nevada to constructive engagement and cooperation; and

WHEREAS Nye County intends to work cooperatively with communities along the Caliente route, the Department of Energy, and any other appropriate group for the purpose of achieving this vision.

NOW THEREFORE, BE IT RESOLVED that Nye County intends to fully, constructively and energetically support:

1. Development of a safe repository at Yucca Mountain,
2. Development of policy that empowers the County concerning repository and transportation safety and health,
3. Creation of synergistic scientific, engineering, education and entrepreneurial economic opportunities in the County,
4. Assisting the United States of America in fulfilling the commitment to provide a geologic repository for spent nuclear fuel and high-level waste to protect the health, safety and welfare of the citizens of the United States,
5. Assisting the United States Department of Energy in meeting their timeline for the reception of spent nuclear fuel and high-level waste at Yucca Mountain,
6. Maximizing jobs and economic opportunities for Nye County citizens,
7. Working cooperatively with appropriate federal entities, rural Nevada communities along the transportation route and other parties willing to constructively engage in the development of a repository that is safe and offers significant economic benefit to Nye County and others most affected by the operation of a repository and related transportation systems.

APPROVED this 20th day of July 2004.

Source: Nye County Board of Commissioners. [Online press release; retrieved 7/30/04.] http://www.nei.org/FileFolder/yucca_nye_county_resolution_7-20-04.pdf.

Nuclear Age Peace Foundation's Top Ten Reasons to Oppose the DoE's Yucca Mountain Plan, David Krieger and Marissa Zubia, August 23, 2002

1. **Accomplishes No Reasonable Objective.**
   Yucca Mountain does not eliminate on-site storage of nuclear waste. After Yucca Mountain is full, there will still be 44,000 tons of high-level nuclear waste stored on-site at reactors throughout the country. There will also be 77,000 tons of such waste moving around the U.S. over the next 30 years, traveling from one of 131

sites an average of 2,000 miles per shipment to Yucca Mountain. If the purpose of the Yucca Mountain project is to consolidate the wastes, that goal will clearly not be achieved.

2. **Provides Minimal Protection.**
   Yucca Mountain itself only provides a small portion of the "protection" that the proposed site promises. The casks that hold the waste are the actual protection, so why Yucca Mountain at all?
3. **Creates More Nuclear Waste.**
   Shipping the waste off-site will allow for the nuclear reactors to continue creating more waste long after the contracts for those sites were set to expire, thus continuing the cycle of producing extremely dangerous waste that no one knows how to safely dispose of. The nuclear industry has economic incentives for moving the waste off-site from the reactors.
4. **Adverse Effects on Future Generations.**
   The project is a distinct danger to defenseless citizens—not just in this generation, but thousands of generations to come will be affected by this decision. Plutonium-239, for example, has a half-life of 24,400 years, which means that the wastes will remain lethal for some 240,000 years.
5. **Earthquake Danger.**
   Yucca Mountain is directly above an active magma pocket and is the third most seismically active area in the United States, with over 600 earthquakes of magnitude 2.5 or greater on the Richter scale in the last 25 years alone. One such earthquake did over a million dollars worth of damage to the US Department of Energy's own testing facility! The most recent earthquake on July 14, 2002 had a magnitude of 4.4.
6. **Fifty Million People Endangered.**
   Routes will move through 734 counties across the United States. The high-level radioactive waste contained in the casks will endanger 50 million innocent people who live within 3 miles of the proposed shipment routes. Hospitals, schools, businesses, emergency personnel, commuters, travelers and passers-by will also cross paths with the shipments that will move through the country at an average rate exceeding six shipments per day. Community health facilities are not adequately prepared or equipped to deal with mass exposure to radioactive matter. To find out how close your residence or place of work is to the proposed routes, enter your address at www.mapscience.org.
7. **Terrorist Attacks.**
   The proposed shipments to Yucca Mountain would move along predictable routes through 44 states, and many metropolitan

areas such as Atlanta (daily shipments), Chicago (every 15 hours), Denver (every 13 hours), and Salt Lake City (every 7 hours). They would provide tempting targets for terrorists.

8. **Costly Accidents and Limited Liability.**
   For each spill that may occur (one out of every 300 shipments is expected to have an accident) the cost of the clean-up is estimated conservatively at $6 billion. Thanks to Congress passing and repeatedly renewing the Price-Anderson Act, the nuclear industry's liability is limited. Taxpayers will pay the bill for accidents even if they occur on reactor property.
9. **Adverse Impact on Water Sources.**
   Yucca Mountain sits above the only source of drinking water for the residents of Amargosa Valley. The aquifer below Yucca Mountain provides water to Nevada's third largest dairy farm, which supplies milk to some 30 million people on the west coast.
10. **Violates Treaties.**
    Yucca Mountain is located on Native American land, belonging to the Western Shoshone by the treaty of Ruby Valley. The Western Shoshone National Council has declared this land a nuclear free zone and demanded an end to nuclear testing and the dumping of nuclear wastes on their land.

It defies reason that to expect that radioactive wastes will sit for tens of thousands of years undisturbed by unpredictable nature, by vengeful terrorists, or by human or technological errors in the design of the containment structure itself. The problem of what to do with high-level nuclear wastes warrants additional consideration and resources, including investigation of alternatives to Yucca Mountain. As an interim solution, the wastes should be converted to dry-cask storage and remain on-site where they were created.

Source: Krieger, David, and Marissa Zubia. [Online article; retrieved 9/22/07.] http://www.wagingpeace.org/articles/2002/08/23_krieger_yucca-top10.htm.

7

Directory of Organizations, Associations, and Agencies

This chapter includes entries from government agencies, industry and trade associations, and other nongovernmental organizations involved in the broad spectrum of waste management activities. These organizations may focus on a specific type of waste, such as hazardous waste, or on several types of waste that are relevant to a business sector, such as nuclear power plants or pulp and paper operations. The directory covers U.S., Canadian, European, and globalized groups that deal with international waste issues, such as toxic waste trading. Entries exclude international telephone numbers because calling accessibility varies from one country to another. Except for federal and state agencies, each entry is accompanied by a brief description of the group's mission, activities, membership, or other identifying notes. Although the information in this section was fact checked prior to publication, it is important to remember that with the growth of the Internet, Web sites change frequently and suffixes and uniform resource locators (URLs) change often. Similarly, organizations may change offices, area codes, suites, and zip codes without warning, so some detective work may be required if contact information does not appear to be accurate.

U.S. Government Agencies

Numerous agencies deal with various types of waste at the federal level, although primary responsibility rests with the Environmental Protection Agency (EPA). The EPA has regional offices throughout the United States, and all have jurisdiction over solid and hazardous waste. Radioactive waste issues are usually handled by the U.S. Department of Energy, which also deals with biomass waste. Other departments, such as Transportation, deal with waste transfer issues, often in concert with related programs under the Nuclear Regulatory Commission. This brief listing identifies some of the major operations of federal agencies, although every department deals with waste in some way, if only in implementing programs to handle in-house waste.

U.S. Department of Energy
Biomass Program
1000 Independence Avenue, SW
Washington, DC 20585
Telephone: (202) 586-5188
Web site: www.eere.energy.gov/biomass/

Federal energy policy includes increasing the role of biomass technologies that decrease U.S. dependence on foreign oil. Biomass is considered a renewable form of energy that utilizes residues from forests and mills that can be substituted for fossil fuels. Biomass can also be used as fuel additives (such as ethanol) and can help reduce greenhouse gases from the decomposition of unused agricultural waste and wood.

U.S. Department of Energy
Office of Civilian Radioactive Waste Management
1000 Independence Avenue, SW
Washington, DC 20585
Telephone: (202) 586-4251
Web site: www.ymp.gov/

This program was established in 1982 under the Nuclear Waste Policy Act to develop and manage a federal system for disposing of spent nuclear fuel from commercial nuclear reactors and high-level nuclear waste from national defense activities. Its operations include the Yucca Mountain site in Nye County, Nevada, a science and technology program, and a waste acceptance and transportation program headquartered in Washington, D.C.

U.S. Department of Energy
Office of Environmental Management
1000 Independence Avenue, SW
Washington, DC 20585
Telephone: (202) 586-5363
Web site: www.em.doe.gov/

Risk reduction and cleanup of the U.S. nuclear weapons program is the responsibility of this organization within the Department of Energy. The staff operate under a mission completion philosophy, working to close existing radioactive waste sites; secure and store nuclear material in a stable, safe configuration in secure locations to protect national security; and transport and dispose of transuranic and low-level wastes.

U.S. Department of Energy
Waste Isolation Pilot Plant
4021 National Parks Highway
Carlsbad, NM 88220
Telephone: (800) 336-9477
Web site: www.wipp.energy.gov/

The Waste Isolation Pilot Plant (WIPP) is a facility used to dispose of defense-related radioactive waste located in the Chihuahuan Desert in New Mexico. The storage site began its waste and disposal operation in March 1999. The site itself is not open to public tours, but an exhibit at an information center in nearby Carlsbad, called "The WIPP Experience," features informative displays and a documentary about the project.

U.S. Department of Transportation
Federal Motor Carrier Safety Administration
1200 New Jersey Avenue, SE
Washington, DC 20590
Telephone: (800) 832-5660
Web site: www.fmcsa.dot.gov/

Established in 2000 as part of the U.S. Department of Transportation, this administration's primary mission is to reduce incidents involving large trucks and buses. The agency also serves as the repository for information on nonradioactive hazardous material and radioactive material routes.

U.S. Environmental Protection Agency
1200 Pennsylvania Avenue, NW
Washington, DC 20460
Telephone: (202) 272-0167
Web site: www.epa.gov/

The Environmental Protection Agency (EPA) has numerous divisions dealing with the many types of waste products, from municipal solid waste and hazardous waste to recycling and pollution prevention; waste treatment and control; programs implementing the Resource Conservation and Recovery Act; and specialized wastes such as batteries, electronic waste (e-waste), scrap tires, used oil, and medical waste. Specialized programs have been developed in partnership with other agencies, schools, and industries, including the GreenScapes Alliance, National Partnership for Environmental Priorities, and the Schools Chemical Cleanout Campaign. Many activities are handled through regional EPA offices.

U.S. EPA
Region 1 (CT, MA, ME, NH, RI, VT)
1 Congress Street, Suite 1100
Boston, MA 02114-2023
Telephone: (617) 918-1111
Web site: www.epa.gov/region01/

U.S. EPA
Region 2 (NJ, NY, PR, VI)
290 Broadway
New York, NY 10007-1866
Telephone: (212) 637-3660
Web site: www.epa.gov/region02/

U.S. EPA
Region 3 (DC, DE, MD, PA, VA, WV)
1650 Arch Street
Philadelphia, PA 19103-2029
Telephone: (215) 814-5000
Web site: www.epa.gov/region03/

U.S. EPA
Region 4 (AL, FL, GA, KY, MS, NC, SC, TN)

Atlanta Federal Center
61 Forsyth Street, SW
Atlanta, GA 30303-3104
Telephone: (404) 562-9900
Web site: www.epa.gov/region04/

U.S. EPA
Region 5 (IL, IN, MI, OH, WI)
77 West Jackson Boulevard
Chicago, IL 60604-3507
Telephone: (312) 353-2000
Web site: www.epa.gov/region05/

U.S. EPA
Region 6 (AR, LA, NM, OK, TX)
Fountain Place, 12th Floor, Suite 1200
1445 Ross Avenue
Dallas, TX 75202-2733
Telephone: (214) 665-2200
Web site: www.epa.gov/region06/

U.S. EPA
Region 7 (IA, KS, MO, NE)
901 North Fifth Street
Kansas City, KS 66101
Telephone: (913) 551-7003
Web site: www.epa.gov/region07/

U.S. EPA
Region 8 (CO, MT, ND, SD, UT, WY)
1595 Wynkoop Street
Denver, CO 80202-1129
Telephone: (303) 312-6312
Web site: www.epa.gov/region08/

U.S. EPA
Region 9 (AZ, CA, HI, NV)
75 Hawthorne Street
San Francisco, CA 94105
Telephone: (415) 947-8000
Web site: www.epa.gov/region09/

U.S. EPA
Region 10 (AK, ID, OR, WA)
1200 Sixth Avenue
Seattle, WA 98101
Telephone: (206) 553-1200
Web site: www.epa.gov/region10/

U.S. Geological Survey
12201 Sunrise Valley Drive
Reston, VA 20192
Telephone: (703) 648-4000
Web site: www.usgs.gov/

Although it has numerous non-waste-related responsibilities, the U.S. Geological Survey manages an interdisciplinary team of geologists, hydrologists, technicians, and other professionals providing support for the Yucca Mountain project in Nevada. Although the branch is under the direction of the Virginia office of the agency, other staff work on the project in California and Nevada.

U.S. Nuclear Regulatory Commission
11555 Rockville Pike
Rockville, MD 20852-2738
Telephone: (301) 415-7000
Web site: www.nrc.gov/

The Nuclear Regulatory Commission's responsibilities include the regulation of radioactive waste, administered through four regional offices and the states. The agency handles three types of waste: low-level waste, including radioactively contaminated protective clothing, tools, filters, and rags; high-level waste, including used nuclear reactor fuel; and uranium mill tailings that remain after the processing of ores to extract uranium and thorium.

U.S. Nuclear Waste Technical Review Board
2300 Clarendon Boulevard, #1300
Arlington, VA 22201
Telephone: (703) 235-4473
Web site: www.nwtrb.gov

This agency was created in 1987 as part of the amendments to the Nuclear Waste Policy Act. Its mission is to review the technical and scientific validity of Department of Energy activities related to the disposal of radioactive waste, including the evaluation of the Yucca Mountain site. It is an independent federal agency composed of 11 part-time members who are eminent in the field of science or engineering, recommended by the National Academy of Sciences.

State Agencies

The management and administration of waste-related activities vary considerably from one state to another. Some states, for instance, place all of the responsibilities under a single "little EPA" that encompasses the entire spectrum of environmental problems and issues. Other states have separate agencies for different types of waste, such as solid waste, radioactive waste, and medical waste. Still another model is to divide responsibility by the type of activity involved, such as finance, compliance, or public education and outreach. Some states also divide environmental management by one of three categories: air, water, and land, which is where waste activities and projects generally are situated from an organizational perspective.

The list below includes the primary agency for waste management in each state; where multiple agencies are involved, the listing reflects the agency with the broadest mandate for waste control, which usually means solid waste. Larger states have several regional offices for waste management, and in these cases, the contact information for the main state office is listed, which is usually at the state capitol. Consider this list a starting point for further research, as waste jurisdictions and organizational structures are subject to change and sometimes overlap. This guidance is important in understanding the waste hierarchy, because only a small handful of states, such as California, approach the problem from the perspective of integrated waste management.

Alabama
Department of Environmental Management
1400 Coliseum Boulevard
Montgomery, AL 36110-2059
Telephone: (334) 271-7700
Web site: www.adem.state.al.us/

Alaska
Department of Environmental Conservation
410 Willoughby Avenue, Suite 303
Juneau, AK 99811-1800
Telephone: (907) 465-5066
Web site: www.dec.state.ak.us/

Arizona
Department of Environmental Quality
1110 West Washington Street
Phoenix, AZ 85007
Telephone: (602) 771-2300
Web site: www.azdeq.gov/

Arkansas
Department of Environmental Quality
8001 National Drive
Little Rock, AR 72209
Telephone: (501) 682-0744
Web site: www.adeq.state.ar.us/

California
Integrated Waste Management Board
1001 I Street
Sacramento, CA 95812-2815
Telephone: (916) 341-6000
Web site: www.ciwmb.ca.gov/

Colorado
Department of Public Health and Environment
700 South Ash Street, Building B
Denver, CO 80246
Telephone: (303) 692-3300
Web site: www.cdphe.state.co.us/

Connecticut
Department of Environmental Protection
Materials and Waste Management
79 Elm Street
Hartford, CT 06106-5127
Telephone: (860) 424-3000
Web site: www.ct.gov/dep/cwp/

Delaware
Department of Natural Resources and Environmental Control
Division of Air and Waste Management
89 Kings Highway
Dover, DE 19901
Telephone: (302) 739-9403
Web site: www.awm.delaware.gov/

Florida
Department of Environmental Protection
Division of Waste Management
2600 Blair Stone Road
Tallahassee, FL 32399-2400
Telephone: (850) 245-8705
Web site: www.dep.state.fl.us/waste/

Georgia
Department of Natural Resources
Environmental Protection Division
2 Martin Luther King Drive
Atlanta, GA 30334
Telephone: (404) 656-7802
Web site: www.gaepd.org/

Hawaii
Department of Health
Solid and Hazardous Waste Branch
919 Ala Moana Boulevard, #212
Honolulu, HI 96814
Telephone: (808) 586-4226
Web site: www.hawaii.gov/health/environmental/waste/

Idaho
Department of Environmental Quality
Waste Management and Remediation Division
1410 North Hilton
Boise, ID 83706
Telephone: (208) 373-0502
Web site: www.deq.idaho.gov/

Illinois
Environmental Protection Agency

Bureau of Land, Waste Management Programs
1021 North Grand Avenue East
Springfield, IL 62794-9276
Telephone: (217) 782-3397
Web site: www.epa.state.il.us/land/waste-mgmt/

Indiana
Department of Environmental Management
100 North Senate Avenue
Indianapolis, IN 46204-2251
Telephone: (800) 451-6027
Web site: www.in.gov/idem/

Iowa
Department of Natural Resources
Waste Management
502 East Ninth Street
Des Moines, IA 50319-0034
Telephone: (515) 281-5918
Web site: www.iowadnr.com/waste/

Kansas
Bureau of Waste Management
1000 SW Jackson Street, Suite 320
Topeka, KS 66612-1366
Telephone: (785) 296-1600
Web site: www.kdheks.gov/waste/

Kentucky
Division of Waste Management
14 Reilly Road
Frankfort, KY 40601
Telephone: (502) 564-6716
Web site: www.waste.ky.gov/

Louisiana
Department of Environmental Quality
P.O. Box 4301
Baton Rouge, LA 70821-4301
Telephone: (225) 219-5337
Web site: www.deq.louisiana.gov/

Maine
Department of Environmental Protection
Bureau of Remediation and Waste Management
17 State House Station
Augusta, ME 04333-0017
Telephone: (207) 287-7688
Web site: www.maine.gov/dep/rwm/

Maryland
Department of the Environment
1800 Washington Boulevard
Baltimore, MD 21230
Telephone: (410) 537-3000
Web site: www.mde.state.md.us/

Massachusetts
Department of Environmental Protection
Waste and Recycling
One Winter Street
Boston, MA 02108
Telephone: (617) 292-5500
Web site: www.mass.gov/dep/

Michigan
Department of Environmental Quality
Waste and Hazardous Materials Division
P.O. Box 30241
Lansing, MI 48909-7741
Telephone: (517) 335-2690
Web site: www.michigan.gov/deq/

Minnesota
Pollution Control Agency
520 Lafayette Road
St. Paul, MN 55155-4194
Telephone: (651) 296-6300
Web site: www.pca.state.mn.us/waste/

Mississippi
Office of Pollution Control
Solid Waste Policy, Planning and Grants Branch

515 Amite Street
Jackson, MS 39201
Telephone: (601) 961-5171
Web site: http://deq.state.ms.us/MDEQ.nsf/page/epd_SolidWasteandMining

Missouri
Department of Natural Resources
Division of Environmental Quality
P.O. Box 176
Jefferson City, MO 65102
Telephone: (573) 751-5401
Web site: www.dnr.mo.gov/env/

Montana
Department of Environmental Quality
1520 East Sixth Street
Helena, MT 59620-0901
Telephone: (406) 444-5300
Web site: www.deq.state.mt.us/

Nebraska
Department of Environmental Quality
1200 N Street, #400
Lincoln, NE 68504
Telephone: (402) 471-2186
Web site: www.deq.state.ne.us/

Nevada
Division of Environmental Protection
Bureau of Waste Management
901 South Stewart Street, #4001
Carson City, NV 89701-5249
Telephone: (775) 687-4670
Web site: http://ndep.nv.gov

New Hampshire
Department of Environmental Services
29 Hazen Drive
Concord, NH 03302-0095
Telephone: (603) 271-3503
Web site: www.des.state.nh.us/

New Jersey
Department of Environmental Protection
Bureau of Solid and Hazardous Waste
401 East State Street
Trenton, NJ 08625
Telephone: (609) 633-1418
Web site: www.state.nj.us/dep/dshw/

New Mexico
Environment Department
1190 St. Francis Drive
Santa Fe, NM 87502
Telephone: (505) 827-0197
Web site: www.nmenv.state.nm.us/

New York
Department of Environmental Conservation
Division of Solid and Hazardous Materials
625 Broadway
Albany, NY 12233-1010
Telephone: (518) 402-8549
Web site: www.dec.ny.gov/

North Carolina
Department of Environment and Natural Resources
Division of Waste Management
1646 Mail Service Center
Raleigh, NC 27699-1646
Telephone: (919) 508-8400
Web site: www.wastenotnc.org/

North Dakota
Department of Health
Division of Waste Management
918 East Divide Avenue, Third Floor
Bismarck, ND 58501-1947
Telephone: (701) 328-5166
Web site: www.health.state.nd.us/wm/

Ohio
Environmental Protection Agency
50 West Town Street, Suite 700

Columbus, OH 43215
Telephone: (614) 644-3020
Web site: www.epa.state.oh.us/

Oklahoma
Land Protection Division
P.O. Box 1677
Oklahoma City, OK 73101-1677
Telephone: (405) 702-5101
Web site: www.deq.state.ok.us/

Oregon
Department of Environmental Quality
Land Quality Division
811 S.W. Sixth Avenue
Portland, OR 97204-1390
Telephone: (503) 229-5696
Web site: www.oregon.gov/DEQ/LQ/

Pennsylvania
Department of Environmental Protection
Bureau of Waste Management
P.O. Box 8471
Harrisburg, PA 17105-8471
Telephone: (717) 783-2388
Web site: www.depweb.state.pa.us/landrecwaste/

Rhode Island
Department of Environmental Management
Office of Waste Management
235 Promenade Street
Providence, RI 02908-2797
Telephone: (401) 222-2797
Web site: www.dem.ri.gov/programs/benviron/waste/

South Carolina
Environmental Quality Control
Bureau of Land and Waste Management
2600 Bull Street
Columbia, SC 29201
Telephone: (803) 896-4000
Web site: www.scdhec.net/lwm/

South Dakota
Department of Environment and Natural Resources
Waste Management Program
523 East Capitol
Pierre, SD 57501
Telephone: (605) 773-3153
Web site: www.state.sd.us/DENR/

Tennessee
Department of Environment and Conservation
Solid and Hazardous Waste Management
401 Church Street, First Floor
Nashville, TN 37243-0435
Telephone: (615) 532-0109
Web site: www.state.tn.us/environment/

Texas
Commission on Environmental Quality
12100 Park 35 Circle
Austin, TX 78753
Telephone: (512) 239-1000
Web site: www.tceq.state.tx.us/

U.S. Virgin Islands
Division of Environmental Protection
Department of Planning and Natural Resources
8100 Lindberg Bay, Suite 61
St. Thomas, U.S. Virgin Islands 00802
Telephone: (340) 774-3320
Web site: www.dpnr.gov.vi/dep/

Utah
Division of Solid and Hazardous Waste
P.O. Box 144880
Salt Lake City, UT 84114-4880
Telephone: (801) 538-6170
Web site: www.hazardouswaste.utah.gov/

Vermont
Waste Management Division
103 South Main Street
Waterbury, VT 05671-0404

Telephone: (802) 241-3888
Web site: www.anr.state.vt.us/dec/

Virginia
Department of Environmental Quality
629 East Main Street
Richmond, VA 23218
Telephone: (804) 698-4000
Web site: www.deq.state.va.us/

Washington
Department of Ecology
P.O. Box 47600
Olympia, WA 98504-7600
Telephone: (360) 407-6000
Web site: www.ecy.wa.gov/

West Virginia
Division of Water and Waste Management
601 57th Street
Charleston, WV 25304
Telephone: (304) 926-0495
Web site: www.wvdep.org/

Wisconsin
Department of Natural Resources
Waste and Materials Management Program
P.O. Box 7921
Madison, WI 53707-7921
Telephone: (608) 266-2621
Web site: www.dnr.state.wi.us/

Wyoming
Department of Environmental Quality
Solid and Hazardous Waste Division
122 West 25th Street
Cheyenne, WY 82002
Telephone: (307) 777-7937
Web site: http://deq.state.wy.us/shwd/

Industry and Trade Associations and Nongovernmental Organizations

Waste generation is common to every business sector, whether the major component of the waste stream is solid waste that can easily be recycled, hazardous waste that requires special handling and treatment, or radioactive waste subject to strict federal rules for disposal. This is why only a sampling of major industry organizations is presented in this directory. Some trade associations have a task force or committee to deal with industry-specific waste issues, such as the Synthetic Organic Chemical Manufacturers Association. Nongovernmental organizations in this list include those that advocate on issues related to specific types of waste processing such as composting, and others that seek the elimination of waste altogether. Among the most active groups are those seeking the end of hazardous waste trading and incineration and enhanced recycling.

100% Recycled Paperboard Alliance
1331 F Street NW, Suite 800
Washington, DC 20004
Telephone: (202) 347-8000
Web site: www.rpa100.com/

The alliance serves as an information resource on the benefits of recycled paperboard and licenses products for labeling. It is an independent, nonprofit organization that actively promotes 100 percent recycled paperboard as a comparable alternative to virgin papers and promotes companies' conversion to these products.

Air & Waste Management Association
One Gateway Center, Third Floor
420 Fort Duquesne Boulevard
Pittsburgh, PA 15222-1435
Telephone: (412) 232-3444
Web site: www.awma.org/

As a professional development organization, this group, founded in 1907, serves to enhance knowledge and expertise by providing a neutral forum for the exchange of information, networking opportunities, public education, international outreach to environmental professionals, and the promotion of global environmental responsibility. The organization's membership includes more than 9,000 environmental professionals in 65 countries, making it one of the most influential organizations in the field. Its flagship publication, *The Journal of the Air & Waste Management Association,* in print since 1951, is the oldest continually published, peer reviewed, technical environmental journal in the world. Another publication, *EM,* is targeted at environmental managers and focuses on regulatory and government policies and management issues. The association also sponsors specialty conferences on topics such as global climate change and offers continuing education courses. One of its current efforts is to build an international network of environmental professionals, and it has established sections in Brazil, Europe, Hong Kong, Mexico, the Philippines, Taiwan, and Saudi Arabia.

Alliance of Foam Packaging Recyclers
1298 Cronson Boulevard, Suite 201
Crofton, MD 21114
Telephone: (410) 451-8340
Web site: www.epspackaging.org/

The alliance works with manufacturers and recyclers of expanded polystyrene (EPS) foam packaging, a light-weight material that is often thrown away. EPS recycling usually focuses on commercial waste generators and customers within a 100- to 200-mile radius. The original materials (identified as no. 6 plastic resin) are collected, then used to make new foam packaging or re-pelletized and remanufactured into products such as plastic lumber. Postconsumer EPS is currently recycled at the rate of 10 to 12 percent each year.

Aluminum Association
1525 Wilson Boulevard, Suite 600
Arlington, VA 22209
Telephone: (703) 358-2960
Web site: www.aluminum.org/

The 200 aluminum plants represented by this group form a trade association for primary producers, recyclers, and makers of semi-fabricated aluminum products, along with companies that provide supplies for the industry. It has offices in Washington, D.C., and Detroit and is governed by a board of directors. Numerous committees provide the structure for the organization, including committees focused on the environment and recycling. The association seeks to strengthen aluminum's market position in comparison with competitive materials and provides research and education to actively address community and environmental concerns.

American Forest & Paper Association
1111 19th Street NW, Suite 800
Washington, DC 20036
Telephone: (800) 878-8878
Web site: www.afandpa.org/

Founded in 1993 as a result of the merger of the National Forest Products Association and the American Paper Institute, the association is the national voice of the forest, pulp, paper, paperboard, and wood products industry. Member companies represent the manufacturers of more than 75 percent of the products produced in the United States. One of the group's strengths is that its members represent the entire spectrum of producers, including small, nonindustrial, private landowners; family-run mills; and large, multiproduct producers. Its mission is to influence successfully public policy that benefits the U.S. paper and forest products industry. The group notes that it has a "big tent" philosophy of inclusion to unite various sectors of the forest products industry.

Appliance Recycling Information Center
1111 19th Street NW, Suite 402
Washington, DC 20036
Telephone: (202) 872-5955
Web site: www.aham.org/

The Center was established in 1993 as a project of the association of Home Appliance Manufacturers to serve as an information clearinghouse on the environmentally responsible disposal and recycling of appliances. Its focus is on industry coordination, in conjunction with the Major Appliance Resource Management Alliance, and on information and education. In 2005, the group began conducting research into major appliance and portable/floor care appliance recycling at the end of product life, typically 10 to 18 years. Discarded appliances are second only to old automobiles as a source of recycled metals, particularly steel.

Asphalt Recycling and Reclaiming Association
#3 Church Circle, Suite PMB 250
Annapolis, MD 21401
Telephone: (410) 267-0023
Web site: www.arra.org/

Because of the continuing need for infrastructure development and road building, this trade association represents an ongoing resource. Members work with the Federal Highway Administration in developing an official policy on recycling and have been responsible for keeping millions of tons of asphalt out of North American landfills. They are also actively involved with the Foundation for Pavement Preservation and the National Center for Pavement Preservation. A cornerstone of the group's efforts is to increase the market share for recycled asphalt.

Association of Postconsumer Plastic Recyclers
2000 L Street NW, Suite 835
Washington, DC 20006
Telephone: (202) 316-3046
Web site: www.plasticsrecycling.org/

This national trade association represents companies that acquire, reprocess, and sell more than 90 percent of North America's postconsumer plastic. It seeks to expand the industry by developing protocols for the design of packaging so it can be more easily recycled, improving the quality of postconsumer plastics entering the system and promoting a cooperative testing program for new packaging. It establishes and updates guidelines for the design of recyclability and standards for good recycling practices. The group's operations include its Technical Committee, which establishes and updates guidelines for the design of recycling practices for recyclability; the Market Development Committee, which actively promotes postconsumer plastic collection and recycled content to end-use markets; and coalitions with other industry organizations such as the American Plastics Council and the National Recycling Coalition.

Association of State and Territorial Solid Waste
Management Officials
444 North Capitol Street NW, Suite 315
Washington, DC 20001

Telephone: (202) 624-5828
Web site: www.astswmo.org/

A 14-member board of directors governs this organization, incorporated in 1974 to focus on the needs of state and territorial solid waste managers who deal with hazardous waste programs, municipal solid waste and industrial waste, recycling and waste minimization, Superfund and state cleanup sites, and underground storage tanks. It has maintained a Washington, D.C., office since 1980 and has a staff of seven. The group holds two membership meetings each year and has a long list of services available to members. Nearly 30 task forces review issues and topics of interest to the association's members, including hazardous waste, underground storage tanks, training and information exchange, and pollution prevention. The association's axiom is, "States learn best from other States."

Basel Action Network
122 South Jackson, Suite 320
Seattle, WA 98104
Telephone: (206) 652-5555
Web site: www.ban.org/

To combat the rising level of trade in toxic waste, the Basel Action Network (BAN) opposes the export of toxic products and technologies from rich to poorer countries. Although the group is not officially connected to the multilateral Basel Convention, it does serve as an information clearinghouse, advocates for international toxic waste policy, and conducts field investigations and campaigns in conjunction with nongovernmental organizations. It works at both the domestic and international level, with a particular focus on Europe (because of its strong leadership in global environmental initiatives), Asia (the primary "victim area" for toxic trade), and the United States (which the group says has a poor record of global stewardship). BAN is recognized by the United Nations Environment Programme as a leading organization dealing with toxic trade issues, and it also works with other groups such as the Organisation for Economic Co-operation and Development. Currently, its campaigns focus on e-waste, green shipbreaking, zero mercury pollution, and ratification of the Basel Ban Amendment.

Battery Council International
401 North Michigan Avenue, 24th Floor
Chicago, IL 60611-4267
Telephone: (312) 644-6610
Web site: www.batterycouncil.org/

Founded in 1924, the council is the trade association for the lead acid battery industry, with more than 175 members worldwide. It is a not-for-profit organization representing businesses that are involved in the lead acid battery life cycle, including manufacturers and recyclers, marketers and retailers, suppliers of raw materials and equipment, and industry consultants. Among its accomplishments are the development of model lead acid battery recycling legislation and a record of recycling 97 percent of all battery lead.

Biodegradable Products Institute
331 West 57th Street, Suite 415
New York, NY 10019
Telephone: (888) 274-5646
Web site: www.bpiworld.org/

This multistakeholder organization represents the interests of government agencies, industry, and university researchers to promote the use of biodegradable and compostable materials. Products that meet specific standards based on at least eight years of research in an approved laboratory are awarded the Compostable logo. The group's members work on a global level, seeking the harmonization of standards for biodegradable products. It also works to further the use and recovery of biodegradable materials.

Blacksmith Institute
2014 Fifth Avenue
New York, NY 10035
Telephone: (646) 742-0200
Web site: www.blacksmithinstitute.org/

Using the metaphor of a blacksmith who makes practical items in a dirty environment, this nonprofit organization was founded in 1999 to assist in pollution remediation. The group is known for its creation of a highly polluted places list and for supporting local partners with financial assistance and grants, technical research, strategic assistance, and networking capability development assistance. The emphasis is on cleaning up toxic waste and contamination where the pollution is ongoing and complex. It works with local nongovernmental organizations to raise awareness about pollution, to create a sound knowledge based on environmental quality, and to strengthen legislation and legal frameworks to curb polluting practices. The primary method for accomplishing this mission is putting resources in the hands of local leaders to build capacity in the developing world. The process begins by identifying polluted places, assessing the health risks at the site, conducting site assessments to enable the design of an intervention, and designing and implementing a remediation strategy. Some projects involve small-scale cleanups, while others deal with ongoing pollution, usually from a specific industrial activity.

Bureau of International Recycling
24 Avenue Franklin Roosevelt
1050 Brussels
Belgium
Web site: www.bir.org/

An international trade federation representing 600 companies from more than 60 countries, the bureau promotes recycling to conserve natural resources. Members are encouraged to do business together, to learn about the most recent advances in technology and market developments, and to learn about international legislative developments. The bureau deals primarily with representatives from the ferrous and nonferrous metals industries, paper, and textiles. Some members also study and trade plastic, rubber, and scrap tires. A new emphasis for the group is China, one of the biggest markets for the association's members.

Can Manufacturers Institute
1730 Rhode Island Avenue NW, Suite 1000
Washington, DC 20036
Telephone: (202) 232-4677
Web site: www.cancentral.com/

The association was initially chartered in 1938 with a membership of 39 manufacturers and suppliers; its members now account for more than 81 percent of annual production of 133 billion cans and employ 22,000 people in 33 states, Puerto Rico, and American Samoa. Its mission is to foster the prosperity of the industry by promoting the can and communicating its benefits to customers, consumers, the media, and trade analysts. It promotes community recycling, sponsors recycling events, and distributes educational materials about environmental awareness.

Cefic
Avenue E. Van Nieuwenhuyse, 4 Box 1
B-1160 Brussels
Belgium
Web site: www.cefic.org/

Incorporated in 1972, Cefic is an international association that represents 29,000 chemical companies that employ about 1.3 million people, accounting for nearly a third of the world's chemical production. It represents European chemical companies, European national federations, and sectoral businesses with a production base in Europe. The group works toward sustainable development and has programs dealing with health, safety, and the environment; international trade and economics; education; transport; and logistics.

Center for Health, Environment and Justice
P.O. Box 6806
Falls Church, VA 22040
Telephone: (703) 237-2249
Web site: www.chej.org/

Love Canal activist Lois Gibbs founded this organization in 1981, originally called the Citizens Clearinghouse for Hazardous Waste. Now the group's name has changed and its agenda has broadened to include all aspects of environmental justice through community organizing and empowerment. The group works with local organizations in an attempt to bring industry and government to work toward an environmentally sustainable future.

Chartered Institution of Wastes Management
9 Saxon Court
St. Peter's Gardens
Marefair

Northampton NN1 1SX
United Kingdom
Web site: www.iwm.co.uk

This professional body sponsors the world's largest annual event devoted to waste management, with participants representing its more than 7,000 members in the United Kingdom and abroad. It operates 10 regional offices to advance the scientific, technical, and practical aspects of waste management to safeguard the environment. The group has a commercial subsidiary, IWM Business Services Limited, that provides support for conferences, exhibitions, training, and technical publications.

Composting Association
3 Burystead Place
Wellingborough
Northampstonshire NN8 1AH
United Kingdom
Web site: www.compost.org.uk/

The association works on behalf of its members to promote the sustainable management of biodegradable resources. It promotes the benefits of composting and other biological treatment techniques and represents its members' views before policy makers. The group's core principles are the development of a diverse, competitive, innovative, profitable industry that promotes good practice and sound environmental management. Members support dialogue with local communities and government authorities to promote a sustainable, financially viable marketplace.

Composting Council of Canada
16 rue Northumberland Street
Toronto, ON M6H 1P7
Canada
Web site: www.compost.org/

The council is a national, nonprofit organization that advocates for composting and compost usage in Canada. It serves as an information clearinghouse and provides networking opportunities for its members. Among the group's initiatives is an effort to develop a national standard for compostable plastic bags to set the criteria for product performance.

Container Recycling Institute
1776 Massachusetts Avenue NW, Suite 800
Washington, DC 20036-1904
Telephone: (202) 263-0999
Web site: www.container-recycling.org/

Founded in 1991, this nonprofit organization focuses on educating government leaders, elected officials, and members of the public on the impact of beverage containers. Its goal is to shift the social and environmental costs of manufacturing, recycling, and disposing of bottles and packaging waste from government and taxpayers to producers and consumers. The group's Web site provides a guide to no-deposit, no-return beverage containers and bottle bill legislation.

Dangerous Goods Advisory Council
1100 H Street NW, Suite 740
Washington, DC 20005
Telephone: (202) 289-4550
Web site: www.hmac.org/

This organization, also known as the Hazardous Materials Advisory Council, was incorporated in 1978. It is an international, nonprofit educational organization with a goal of promoting safety in domestic and international transportation of dangerous goods. Members include shippers, carriers, container manufacturers and reconditioners, emergency and waste cleanup companies, and trade associations. It works with the U.S. Department of Transportation and regulatory counterparts in Mexico and Canada.

Electronic Industry Alliance
2500 Wilson Boulevard
Arlington, VA 22201
Telephone: (703) 907-7500
Web site: www.eia.org/

The alliance represents U.S. electronics manufacturers as a partnership of associations and 1,300 companies. Its organization includes an environmental issues council that assists members in dealing with national and international policies and regulations affecting the industry, including those related to electronic waste. The organization is directed by a board of governors that brings together the common interests of various sectors of the industry while preserving the autonomy of member companies.

Environmental Services Association
154 Buckingham Palace Road
London SW1W 9TR
United Kingdom
Web site: www.esauk.org/

Waste management and secondary resources industries throughout the United Kingdom are represented through this London-based association. Representing both public- and private-sector agencies and companies, the group works with the government and regulators to bring a sustainable system of waste management to the United Kingdom. The organization deals with issues ranging from climate change and energy to the development of waste protocols, composting, and sustainability.

Environmental Working Group
1436 U Street NW, Suite 100
Washington, DC 20009
Telephone: (202) 667-6982
Web site: www.ewg.org/

The Environmental Working Group uses a team approach utilizing scientists, engineers, policy experts, lawyers, and computer programmers to analyze legal documents, scientific studies, government data, and laboratory tests to review health and environmental risks. Its work is divided into three categories: health and toxics, farming, and natural resources. It has produced reports on the transport of nuclear waste and the environmental impact of hazardous waste.

European Compost Network
Postbox 22 29
D-99403 Weimar
Germany
Web site: www.compostnetwork.info

This collaborative group promotes sustainable practices in composting, anaerobic digestion, and other treatment procedures for organic residues across Europe. It operates through working groups for the development of a common European strategy for biowaste, standards and quality monitoring, composting marketing and application, the development of industry practices, and support for Mediterranean and Eastern European countries. The organization is a collaborative and is part of the ORBIT Association, a not-for-profit organization that promotes the scientific development of environmental technology worldwide.

Fibre Box Association
25 Northwest Point Boulevard, Suite 510
Elk Grove Village, IL 60007
Telephone: (847) 364-9600
Web site: www.fibrebox.org/

This nonprofit organization serves North American manufacturers of corrugated containers. It compiles and disseminates statistical information, represents members' interests before government and regulatory boards, monitors regulation and legislation, and supports the safety of industry employees. It also promotes the container market at both the national and international levels and seeks to advance technical developments for the industry.

Film and Bag Federation
1667 K Street NW, Suite 1000
Washington, DC 20006
Telephone: (202) 974-5215
Web site: www.plasticbag.com/

As part of the Society of the Plastics Industry, the federation serves as the voice of plastic film and bag manufacturers and develops positions on regulatory and legislative issues related to energy, health, worker safety, recycling, and the environment. It works with its counterparts in other countries such as Australia and provides services to its members. The organization's goal is to increase its membership to represent a greater proportion of the industry segment and to provide an inclusive forum to discuss and take action on issues important to the film and bag industry.

Glass Packaging Institute
700 North Fairfax Street, Suite 510
Alexandria, VA 22314

Telephone: (703) 684-6359
Web site: www.gpi.org/

North American and Mexican glass container manufacturers are represented by this trade association, which advocates industry standards, promotes sound environmental policies, and educates packaging professionals. The members produce glass containers for food, beverage, cosmetic, medicine, and other products. Decisions are made by a board of trustees consisting of one representative from each member company and one representative from associate members, closure manufacturers, and foreign-member companies. The Institute's work is also delegated to several committees and issue task forces.

Global Anti-Incinerator Alliance
Unit 320 Eagle Court Condominiums
26 Matalino Street
Quezon City 1101
Philippines
Web site: www.no-burn.org/

Also known as GAIA, this international alliance was formed in 2000 to collaborate and end incineration of waste and to promote safe, economical alternatives to waste management. The organization focuses on three waste streams: municipal discards, hazardous waste, and medical waste (in conjunction with the group Health Care Without Harm). It also works to stop the World Bank from funding incinerators around the world. The group's work is carried out through regional networks and issue workgroups to deal with issues that transcend national and regional borders.

Industry Council for Electronic Equipment Recycling
6 Bath Place
Rivington Street
London EC2A 3JE
United Kingdom
Web site: www.icer.org/

Designed primarily for companies operating in the United Kingdom, this lobbying organization also runs accreditation programs for waste equipment recyclers and refurbishers, carries out research, and provides members with information on regulations and other legislation affecting the industry.

Institute for Recyclable Materials
1419 CEBA
Louisiana State University College of Engineering
Baton Rouge, LA 70803
Telephone: (504) 388-8650
Web site: www.laregents.org/

Operating primarily as a research program, the institute seeks to prevent, minimize, or recycle municipal solid waste through new and improved technologies. The core staff of engineers and scientists conducts multidisciplinary investigation at various sites on the Louisiana State University campus. It also operates the Transcontinental Materials Exchange for waste generators seeking to sell or give away by-product materials.

Institute of Packaging Professionals
1601 Bond Street, Suite 101
Naperville, IL 60563
Telephone: (630) 544-5050
Web site: www.iopp.org/

By providing global leadership in packaging, the group's members offer education and training for packaging professionals for career enhancement and promote the idea that packaging is a positive, environmentally responsible, and economically efficient force in a modern society. The organization is managed by a small staff and volunteers at its Chicago-area headquarters. It provides a speakers' bureau, technical committee assistance, a journal, and a newsletter, and it operates a career center.

Institute of Scrap Recycling Industries
1615 L Street NW, Suite 600
Washington, DC 20036-5610
Telephone: (202) 662-8500
Web site: www.isri.org/

The scrap industry was initially represented by the National Association of Waste Material Dealers and the Institute of Scrap Iron and Steel, which merged to become the Institute of Scrap Recycling Industries because of the overlap of their membership. The institute is organized by chapters representing more than 1,200 companies that process, broker, and consume scrap commodities, providing advocacy, education, and compliance training. Members represent firms engaged in the recycling of metals, paper, plastics, glass, rubber, electronics, and textiles. The group's activities are driven by policy statements on issues such as design for recycling, tax relief, and marketable credits for secondary lead recycling.

Integrated Waste Services Association
1331 H Street NW, Suite 801
Washington, DC 20005
Telephone: (202) 467-6240
Web site: www.wte.org/

Formed in 1991, the association encourages the use of waste-to-energy facilities and technology as part of a comprehensive waste management policy. Waste-to-energy plants serve municipal governments through the combustion of household trash at the 87 facilities now in operation. By advocating integrated solutions to waste management, the members of the group seek to provide an alternative to conventional power plants and disposal in landfills.

International Association of Electronics Recyclers
P.O. Box 16222
Albany, NY 12212-6222
Telephone: (888) 989-4237
Web site: www.iaer.org

As the first trade association for the electronic recyclers industry, established in 1998, this not-for-profit group seeks the development of an effective and efficient infrastructure for managing the life cycle of electronics products. Its members represent a cross-section of industry sectors, including recyclers and industry associations. It is governed by a board of directors responsible for oversight and guidance.

International Bottled Water Association
1700 Diagonal Road, Suite 650
Alexandria, VA 22314
Telephone: (703) 683-5213
Web site: www.bottledwater.org/

With the public's increasing attention on recycling of beverage containers, this trade association is becoming more active in environmental issues. The group was founded in 1958 to represent the interests of U.S. and international bottlers, distributors, and suppliers. Members also include the manufacturers of small equipment; original family-owned and -operated water bottlers; and large, diversified food corporations with bottled water as part of their product lines. One of the organization's activities is the development of a model code, used as a regulation in many states, that addresses requirements for the production and sale of bottled water. The association notes that it is committed to actively participating in recycling and educating the public about the importance of recycling.

International Corrugated Case Association
25 Northwest Point Boulevard, Suite 510
Elk Grove Village, IL 60007
Telephone: (847) 364-9600
Web site: www.iccanet.org/

Formed in 1961, this trade association promotes the general welfare of the world's corrugated container industry by collecting and disseminating information and serving as the global platform for addressing the needs of members that can be more effectively addressed by an association rather than by individual members. It has six work groups and sponsors a biennial management conference. The group is managed by a board of directors and helps create a worldwide network for problem solving, information sharing, and a best practices arena.

International Solid Waste Association
Vesterbrograde 74, Third Floor
DK-1620 Copenhagen
Denmark
Web site: www.iswa.org/

The International Solid Waste Association (ISWA) is an independent, nongovernmental organization whose mission is to expand the exchange of information and expertise on solid waste management on a global level. The ISWA conducts its activities through working groups and a scientific and technical committee examining both public- and private-sector waste management throughout the world. Members include individuals and organizations from the scientific community, public institutions, and public and private companies worldwide that are interested in waste management issues. The group provides members with access to publications and newsletters, conferences and meetings, and the opportunity to work with international organizations such as the United Nations Environment Programme.

Keep America Beautiful
1010 Washington Boulevard
Stamford, CT 06901
Telephone: (203) 323-8987
Web site: www.kab.org/

Keep America Beautiful was founded in 1953 as the United States' largest nonprofit community improvement and educational organization. The group's three areas of focus are litter prevention, beautification and community improvement, and waste reduction. Each year, its members participate in the Great American Cleanup, working to improve the local environment and encourage recycling.

Mid-Atlantic Consortium of Recycling and
Economic Development Officials
1504 South Street
Philadelphia, PA 19146
Web site: www.macredo.org/

This organization brings together recycling and economic development officials from Delaware, Maryland, Pennsylvania, Virginia, Washington, D.C., and West Virginia to identify, promote, and implement projects on a regional basis. Its goal is to stimulate the demand for postconsumer materials; promote economic growth and create jobs; and develop an efficient regional recycling infrastructure by researching recycling market issues, offering a forum for exchange among members, and providing region-wide publications. Unlike local or state-based projects, this group seeks to combine individual and local recycling and job creation efforts to maximize regional success.

Municipal Waste Management Association
1620 Eye Street NW, Suite 300
Washington, DC 20006
Telephone: (202) 861-6775
Web site: www.usmayors.org/uscm/mwma/

The association, formed in 1982, is affiliated with the Washington, D.C.–based U.S. Conference of Mayors. Its goal is to promote operational efficiencies, facilitate information, foster innovation, and promote legislative advocacy dealing with waste management issues. The group focuses on Superfund implementation, brownfields redevelopment, clean air, clean water, and waste-to-energy programs. Members represent municipal solid waste managers and directors, environmental commissioners, and public works professionals.

National Association for PET Container Resources
P.O. Box 1327
Sonoma, CA 95476
Telephone: (707) 996-4207
Web site: www.napcor.com/

Polyethylene terephthalate (PET) plastics are those commonly used in bottling and other packaging industries. The organization was founded in 1987 and serves both the United States and Canada. Its members promote the introduction and use of PET packaging through its life cycle and educates the public and officials about its value as an environmentally sustainable package. Among the group's goals is public education about PET products and packaging and the use of single-serve containers and recycling.

National Association of Manufacturers
1331 Pennsylvania Avenue NW
Washington, DC 20004-1790
Telephone: (202) 637-3000
Web site: www.nam.org/

One of the most powerful trade associations in the United States, the National Association of Manufacturers advocates for the competitiveness of manufacturers by shaping a legislative and regulatory environment conducive to U.S. economic growth. It has a pro-growth, pro-manufacturing agenda and seeks to be the primary source for information on manufacturers' contributions to innovation and productivity. Two of its waste-related projects involve the recycling of electronic products and a manual on waste reduction for members.

National Office Paper Recycling Project
1620 Eye Street NW

Washington, DC 20006
Telephone: (202) 293-7330
Web site: http://www.usmayors.org/uscm/uscm_projects_services/environment/national_paper_recycling_project.html

Launched publicly in September 1990, this project is a long-term collaboration of the U.S. Conference of Mayors with private corporations and public interest groups. This national campaign sponsors efforts to create office paper recycling programs and encourages the purchase of recycled office products. The group provides cities with technical assistance, market information, data, and guidebooks for setting up programs. The group sponsors the National Office Paper Recycling Challenge for city offices and conducts research on the supply of and demand for waste paper and office paper recycling.

National Recycling Coalition
1325 G Street NW, Suite 1025
Washington, DC 20005
Telephone: (202) 347-0450
Web site: www.nrc-recycle.org/

As a charitable, nonprofit organization composed of recycling professionals and advocates, the National Recycling Coalition (NRC) was founded in 1978 to eliminate waste and promote sustainable economies through advancing sound management practices for raw materials in North America. It is often considered the leading national voice on recycling and includes state and regional recycling organizations as affiliates. The NRC's strategic plan includes five goals: being recognized as the leading national voice on recycling, serving as a catalyst for constructive dialogue on sustainable raw material management, advancing members' understanding of current and emerging recycling issues, fostering new and innovative recycling solutions, and maximizing its resources to ensure the group's organizational health and vitality.

National Solid Wastes Management Association
4301 Connecticut Avenue NW, Suite 300
Washington, DC 20008-2304
Telephone: (202) 244-4700
Web site: www.nswma.org/

The association represents companies that provide solid, hazardous, and medical waste collection in North America. The group was founded in 1962 with a mission of promoting the management of waste in an environmentally responsible, efficient, profitable, and ethical manner. Activities include education and training opportunities, research, and federal and state issue advocacy. It is managed through a board of governors, chapter officers, and a women's council.

National Waste Prevention Coalition
201 South Jackson Street, #701
Seattle, WA 98104-3855
Telephone: (206) 296-4481
Web site: www.metrokc.gov/dnrp/swd/nwpc/

Solid waste management professionals who work for state and local governments, nonprofit organizations, universities, and consultants dedicated to the prevention of waste creation and the reduction of the use of resources are members of the coalition, which was founded in 1994. Their projects have included the National Junk Mail Reduction Campaign; the Model Cleaners Project, which recognizes leaders in the dry cleaning and "wet cleaning" industry; and a telephone book reduction program.

North American Hazardous Materials Management Association
11166 Huron Street, Suite 27
Denver, CO 80234
Telephone: (877) 292-1403
Web site: www.nahmma.org/

The association was established in 1993 to reduce the hazardous materials constituents entering the waste stream and to prevent pollution. Members include manufacturers, government regulators, waste-handling businesses, nonprofit environmental organizations, and government officials. The focus is on hazardous materials not otherwise covered by regulations that enter the municipal waste stream from households and small businesses. Activities are directed by a board of directors elected from the membership, which sets policy and is responsible for the work of the corporation.

Northeast Recycling Council
139 Main Street, Suite 401

Brattleboro, VT 05301
Telephone: (802) 254-3636
Web site: www.nerc.org/

The council is one of the largest regional organizations in the United States, promoting packaging waste, recycling, compost and organics management, and green procurement. The 10 Northeast states that are members promote recycling market development through business assistance and legislative advocacy. Specific programs have been developed to deal with issues such as compost and organics management, procuring green electronics, and marine shrink-wrap recycling.

Plastic Loose Fill Council
P.O. Box 21040
Oakland, CA 94620
Telephone: (510) 654-0756
Web site: www.loosefillpackaging.com/

Commonly known as packaging "peanuts," polystyrene loose fill can be recycled and reused. This trade organizations helps consumers find their nearest drop-off center for leftover plastic peanuts, with more than 1,500 centers now in operation for collection. The council operates the Peanut Hotline to inform consumers and answer environmental questions about plastic loose fill recycling.

Plastics Foodservice Packaging Group
1300 Wilson Boulevard
Arlington, VA 22209
Telephone: (703) 741-5647
Web site: www.polystyrene.org/

This group is made up of 10 major resin suppliers and manufacturers of polystyrene and plastic food service products. Working at both the domestic and global levels, the aim is to educate the public on how these items benefit the environment and on the important benefits they provide for health and safety. Formerly known as the Polystyrene Packaging Council, this is a business group of the American Chemistry Council. It is now working with similar groups in Asia, Europe, and North and Latin America to facilitate communication across international boundaries.

Product Stewardship Institute
137 Newbury Street, Seventh Floor
Boston, MA 02116
Telephone: (617) 236-4855
Web site: www.productstewardship.us/

Created in 2000 as part of the first forum where government officials could discuss product stewardship policies, this group initially dealt with five product categories: electronics, paints, mercury-containing products, pesticides, and tires. Product Stewardship Action Plans were created for each product, starting with electronic management, with plans for seven categories now being developed.

Rechargeable Battery Recycling Corporation
1000 Parkwood Circle, Suite 450
Atlanta, GA 30339
Telephone: (678) 419-9990
Web site: www.rbrc.org/

To reduce the number of products entering the waste stream, this organization collects used cellular telephones and portable rechargeable batteries, develops drop-off sites, and conducts educational outreach. It also works with community organizations to find uses for cellular phones rather than disposing of them. The organization operates more than 30,000 drop-off sites in the United States.

Reuse Development Organization
2 North Kresson Street
Baltimore, MD 21224
Telephone: (410) 558-3625
Web site: www.redo.org/

The Reuse Development Organization was established as a nonprofit organization in 1995 after a conference identified participants' desire to form a group whose members advocate the redistribution of materials from one entity that cannot use them to one that can do so. It is one way for suppliers to conserve natural resources, reduce greenhouse gases, and get materials to disadvantaged people and organizations. It provides education, training, and technical assistance to initiate and implement reuse programs.

Sierra Club
85 Second Street, Second Floor
San Francisco, CA 94105
Telephone: (415) 977-5500
Web site: www.sierraclub.org/

Although the Sierra Club is most frequently associated with the protection of wilderness areas, endangered species, and outdoor recreation, it also has a major campaign concerning nuclear waste. The group has developed several policies dealing with issues such as nuclear waste transportation, low-level radioactive waste, radioactive scrap, and problems specific to Colorado.

Silicon Valley Toxics Coalition
760 North First Street
San Jose, CA 95112
Telephone: (408) 287-6707
Web site: www.etoxics.org/

One of the oldest groups working with the high-tech industry, the coalition's members work to shift industries toward greater sustainability. In 1982, the organization responded to the discovery of groundwater contamination in the Silicon Valley of California and advocated legislation to keep the community informed about the identification of Superfund sites and subsequent cleanup activities.

Society of the Plastics Industry
1667 K Street NW, Suite 1000
Washington, DC 20006
Telephone: (202) 974-5200
Web site: www.plasticsindustry.org/

This trade association, founded in 1937, represents the entire plastic industry supply chain, from processing and raw materials suppliers to machinery and equipment manufacturers. The plastics industry employs 1.1 million workers and provides more than $341 billion in annual shipments. The association works through regional offices and offers programs for members on the local, national, and international level. It develops codes and standards, organizes committees around topics such as the environment and international trade, and has programs for outreach and education.

Solid Waste Association of North America
1100 Wayne Avenue, Suite 700
Silver Spring, MD 20910
Telephone: (800) 467-9262
Web site: www.swana.org/

An association of more than 7,600 members throughout North America, the Solid Waste Association of North America (SWANA) has a mission of education, innovation, and communication. Its nearly 50 local chapters provide forums for members, advocate on solid waste policy, provide training, host conferences, and provide certification in the various stages of solid waste management. The organization is governed by an international board and has eight technical sections dealing with issues ranging from waste collection and transfer to landfill management. Training opportunities, including electronic course work, are coupled with event training, home study classes, and on-site training for those seeking careers in solid waste management. Another SWANA associated group is the Applied Research Foundation, which leverages research dollars for issues addressing solid waste concerns.

Steel Recycling Institute
Iderson Drive
Pittsburgh, PA 15220-2700
Telephone: (412) 922-2772
Web site: www.recycle-steel.org/

Established in 1988 as the Steel Can Recycling Institute, this nonprofit trade organization promotes steel can recycling. In 1993, the group expanded its focus to encompass the recycling of all steel products, including automobiles, appliances, and construction materials. Recycled steel can be used to make other steel products, and current steel recycling rates exceed 50 percent. Its activities include a grassroots effort to work directly with the public and private sectors to implement recycling.

Synthetic Organic Chemical Manufacturers Association
1850 M Street NW, Suite 700
Washington, DC 20036-5810
Telephone: (202) 721-4100
Web site: www.socma.com/

The 275 member companies of this association represent specialty, batch, and custom chemical manufacturers at 2,000 sites with more than 100,000 employees. Founded in 1921, the group includes small specialty producers and large multinational corporations. Products include pharmaceuticals and cosmetics, soaps, plastics, industrial products, and other materials using refined chemicals. It is managed by a board of governors and includes an environmental committee dealing with hazardous waste. Members can participate in conferences and training on subjects such as compliance, stewardship, and business development.

Tire Industry Association
1532 Pointer Ridge Place, Suite G
Bowie, MD 20716-1883
Telephone: (301) 430-7280
Web site: www.tireindustry.org/

The association and its members host the Tire and Rubber Recycling Advisory Council, which promotes the reuse and recovery of tires and tire-derived materials. It works to enhance the recycling of tires and recyclable rubber products and promotes the industry's position in the legislative, judicial, and regulatory arenas. More than 4,500 members are in the group, which was formed in July 2002 with the merger of the International Tire and Rubber Association and the Tire Association of North America.

U.S. Composting Council
4250 Veterans Memorial Highway, #275
Holbrook, NY 11741
Telephone: (631) 737-4939
Web site: www.compostingcouncil.org/

This nonprofit organization is a nationally based group committed to the advancement of the composting industry. It has about 200 members who are dedicated to the development, expansion, and promotion of composting through research, promoting best practices, establishing standards, educating the public and professionals, and enhancing product quality and markets. It also directs the Composting Council Research and Education Foundation, a nonprofit charitable foundation that administers public and private research and education grant activities.

WASTE
Nieuwehaven 201
2801 CW Gouda
The Netherlands
Web site: www.waste.nl/

The Netherlands-based nonprofit organization WASTE is active in providing technical assistance in solid waste management and resource recovery, low-cost sanitation and liquid waste management, community-based environmental improvement, and micro- and small enterprise development through local partnerships in developing nations. It is currently active in Peru, Costa Rica, Mali, the Philippines, India, and Bulgaria. The organization has a multidisciplinary team of consultants and experts on urban environment, solid waste management, urban planning, sanitation, and environmental economy based in Gouda, the Netherlands. Short-term assistance is provided to organizations that are responsible for running their own projects. Training, planning, and management support are provided to increase capacity and to strengthen project sustainability.

Waste Equipment Technology Association
4301 Connecticut Avenue NW, Suite 300
Washington, DC 20008-2304
Telephone: (202) 244-4700
Web site: www.wastec.org

Founded in 1972 as the Waste Equipment Manufacturers Institute, and renamed in 1993, this group is a quasi-independent trade association representing companies that design, build, distribute, and service equipment for, or consult with businesses in this field dealing with solid and hazardous waste and recyclable materials. It consists of four program divisions: technical, executive, market enhancement, and education and information. Waste management professionals, manufacturers, distributors, recyclers, consultants, equipment leasing companies, and for-profit companies that provide solid and medical waste collection services are represented by this group, which provides information to its members on industry trends and developments; landfill design, operation, and maintenance; vendors; and safety standards for waste management equipment. Public officials can become affiliate members, and the association also has a membership category for international members that do not have operations in the United States and Canada.

Waste Watch
56–64 Leonard Street
London EC2A 4LT
United Kingdom
Web site: www.wastewatch.org.uk/

Waste Watch is an independent British charity that promotes the sustainable use of resources. Its goal is to increase awareness about waste from the perspective of reducing, reusing, and recycling. Funded by the central government, charitable trusts, corporations, individuals, local government, and the national lottery, the group focuses on changing attitudes through communication, education, information, and research.

Zero Waste America
217 South Jessup Street
Philadelphia, PA 19107
Telephone: (215) 629-3553
Web site: www.zerowasteamerica.org/

Lynn Landes, a journalist, founded Zero Waste America (ZWA), an Internet-based environmental research organization that provides information on legislative, legal, technical, environmental, health, and consumer issues and specializes in waste disposal issues. A nonprofit organization, it focuses attention on the regulatory enforcement of the Resource Conservation and Recovery Act. Landes contends that the United States lacks a federal waste management plan and says the group is "non-profit and unconventional. We accept no money. ZWA has no membership, regular meetings, or conventional office holding." ZWA is not associated with any business or government.

8

Resources

Print Resources

Books and Monographs

Ackerman, Frank. 1996. *Why Do We Recycle? Markets, Values, and Public Policy.* Washington, DC: Island Press. 222 pages.

When municipal governments began massive recycling programs in the 1980s and 1990s, most officials had no idea that there would soon be a drop-off in the market for recyclable goods that would not reappear until the next decade. The author provides pro and con arguments for recycling, with most of the opposing views based on the question of whether recycling is economically feasible. He critiques those who base their decisions primarily on economic issues, rather than environmental ones, and notes that a need exists to think about recycling from ethical and social perspectives as well.

Addis, Bill. 2005. *Building with Reclaimed Components and Materials.* London: Earthscan. 224 pages.

Recycling involves more than just putting paper, cardboard, and aluminum cans in bins at the curb. Green and sustainable design also involves finding ways to recycle components and materials in buildings. This book reviews the ways in which construction practices can be integrated to consider recycled products, using case studies and advice on how to reduce the amount of waste going to landfills.

Allen, Barbara L. 2003. *Uneasy Alchemy: Citizens and Experts in Louisiana's Chemical Corridor Disputes.* Cambridge, MA: MIT Press. 224 pages.

The hazardous waste disputes that are common in Louisiana's petrochemical corridor and Cancer Alley are rooted in a historical context that is the subject of this book. Former plantations are now home to toxic waste, and otherwise valuable lands are near Superfund sites. The author identifies rural communities that are experiencing "environmental injustice" because the land in southern Louisiana is cheap and degraded by huge corporate interests.

Applegate, John S., and Jan G. Laitos. 2005. *Environmental Law: RCRA, CERCLA, and the Management of Hazardous Waste.* New York: Foundation Press. 318 pages.

The authors contend that by understanding hazardous waste law, policy makers can see how science and policy sometimes conflict and how waste management legislation has become increasingly complex. When waste is mismanaged, the resultant liability affects entire communities, as seen in the example of Love Canal, New York.

Bagchi, Amalendu. 2004. *Design, Construction, and Monitoring of Landfills.* Hoboken, NJ: John Wiley. 712 pages.

This book provides an overview of each of the types of waste management used today, from incineration and composting to recycling and reuse. While the focus is on landfills, the issues presented are common to many waste management options, such as site selection, construction, and health and safety problems. The author discusses the role of integrated solid waste management, with new material on medical waste management, wetlands mitigation, and landfill remediation.

Basel Action Network. 2005. *The Digital Dump: Exporting Re-Use and Abuse to Africa.* Seattle: Basel Action Network. Available at www.ban.org.

This is a follow-up report on the problems associated with electronic waste (e-waste). The authors identify the primary exporters of e-waste and the countries to which it is routinely sent. It cautions consumers about the need to research whether a recycling company actually deals with the materials and products or ships them to developing countries where environmental controls are less stringent.

Basel Action Network and Silicon Valley Toxics Coalition. 2002. ***Exporting Harm: The High-Tech Trashing of Asia.*** **Seattle: Basel Action Network. Available at www.ban.org.**

The Basel Action Network joins with the Coalition in this pioneering exposé of the export of electronic waste. The report is coupled with a film by the same name that shows children playing on piles of e-waste with no protection against the chemicals and metallic parts of used computers, monitors, and keyboards.

Bayliss, Colin, and Kevin Langley. 2003. ***Nuclear Decommissioning, Waste Management, and Environmental Site Remediation.*** **Boston: Butterworth Heinemann. 330 pages.**

One of the most vexing problems associated with nuclear power is what to do with the waste created in the operation of nuclear facilities. This somewhat technical book covers global regulators and agreements, such as the International Atomic Energy Agency, the International Commission on Radiological Protection, and the Oslo/Paris Convention. It explains the historical involvement of government and current challenges as older facilities are decommissioned.

Blackman, William C. 2001. ***Basic Hazardous Waste Management.*** **2nd ed. Boca Raton, FL: Lewis Publishers. 468 pages.**

For an extensive analysis of hazardous waste, this volume provides almost everything a reader could ever want, from the definition and generation of waste to treatment and disposal methods to how hazardous waste sites can be cleaned up. Also included are sections on more diverse topics such as phytoremediation, medical and biological wastes, radioactive waste, and worker health issues.

Brummet, Dave, and Lillian Brummet. 2004. ***Trash Talk: An Inspirational Guide to Saving Time and Money through Better Waste and Resource Management.*** **Frederick, MD: Publish America. 192 pages.**

By explaining how North Americans account for 8 percent of the world's population and 50 percent of its garbage, the authors have compiled a manual to change those percentages. The book identifies ways that household trash can be reused with simple strategies that change our habits about throwing things away.

Buclet, Nicholas, ed. 2007. *Municipal Waste Management in Europe*. Berlin: Springer. 216 pages.

European waste management is viewed not so much as a technical matter but as a political and social one. The book is targeted toward those interested in economics and politics rather than science, concluding that a real need exists for standardization and harmonization among municipal entities.

Bullard, Robert D. 2000. *Dumping in Dixie: Race, Class, and Environmental Quality*. Boulder, CO: Westview. 234 pages.

This classic explanation of environmental justice is written by an environmental sociologist who examines the widening socioeconomic disparities that create environmental discrimination, best viewed through the lens of the siting of municipal landfills, incinerators, hazardous waste facilities, and other noxious land uses in African American communities.

Burns, Loree Griffin. 2007. *Tracking Trash: Flotsam, Jetsam, and the Science of Ocean Motion*. Boston: Houghton Mifflin. 58 pages.

This book provides a short, simple introduction to understanding how human debris is moved through ocean currents and ends up on beaches. The author profiles oceanographer Dr. Curtis Ebbesmeyer, who tracks trash and looks for ways to preserve the marine environment and habitats.

Cairney, T., and D. M. Hobson, eds. 1998. *Contaminated Land: Problems and Solutions*, 2nd ed. London: Routledge. 369 pages.

The cover of this book depicts a skull and crossbones, indicative of the types of issues raised in this edited volume. The authors note that when hazardous waste was first identified as a problem in the late 1970s, very little was known about the risks, and therefore, little concern was raised about what to do with it, especially in comparison with water and air pollution. Although this book is more technically oriented than others, it does provide some background on the response by the United Kingdom and the

Netherlands, with limited commentary on other European countries and sections on the United States.

Chapman, Neil A. 2003. ***Principles and Standards for the Disposal of Long-Lived Radioactive Wastes.*** **Boston: Pergamon. 277 pages.**

Part of a series on waste management issues, this book is more technical than most readers might need, but it does help to illustrate the complexity of dealing with radioactive waste. In-ground disposal is suggested as the most desirable method for dealing with waste created through energy production, although it is clear that this is not the only method that should be considered by officials.

Cheremisinoff, Nicholas P. 2003. ***Handbook of Solid Waste Management and Waste Minimization Technologies.*** **Boston: Butterworth-Heinemann. 477 pages.**

Unlike some books that focus only on technology, this volume is notable for some limited coverage of the policy drivers and environmental laws that regulate solid waste. Large sections of the book cover biosolids and waste minimization methods, but most of the book deals with the technology of operating a landfill and handling municipal solid waste.

Church, Thomas W. 1993. ***Cleaning Up the Mess: Implementation Strategies in Superfund.*** **Washington, DC: Brookings. 209 pages.**

When Superfund legislation was initially enacted by Congress, it was assumed that it would be a panacea for all of the country's hazardous waste site cleanups. The case studies in this early analysis of the program's operation show, however, that establishing liability for cleanup has always been a problem that equals the costs associated with the pollution itself.

Clapp, Jennifer. 2001. ***Toxic Exports: The Transfer of Hazardous Wastes from Rich to Poor Countries.*** **Ithaca, NY: Cornell University Press. 178 pages.**

The author exposes how international firms in industrialized countries have routinely engaged in shipping hazardous waste to poor countries that are ill equipped to deal with the materials. She argues that the 1989 Basel Convention is too weak to control the waste trade and notes that the practice is too lucrative for either business interests or poor countries to stop doing it. One example is the use of labels identifying materials destined for "recycling" as a way to circumvent regulations that prohibit the transfer of hazardous waste.

Colten, Craig E., and Peter N. Skinner. 1996. ***The Road to Love Canal: Managing Industrial Waste before EPA.*** **Austin: University of Texas Press. 217 pages.**

The environmental disaster known as Love Canal is a story about a hazardous waste dump that operated before much concern was raised about industrial waste practices. Prior to 1970, waste producers commonly dumped chemicals and other toxic substances into local waterways and open pits. These practices were considered acceptable and even recommended after World War II, with little thought as to any adverse consequences. Contemporary debates and litigation over liability are complicated by the fact that almost no regulations were in effect at the time.

Commission for Racial Justice. 1987. ***Toxic Wastes and Race in the United States: A National Report on the Racial and Socioeconomic Characteristics of Communities with Hazardous Waste Sites.*** **New York: United Church of Christ.**

The Commission's landmark study heightened public awareness of the siting of hazardous waste facilities and other locally unwanted land uses in neighborhoods. The study determined that race was the most potent variable in predicting where commercial hazardous waste facilities were located in the United States.

Crooks, Harold. 1993. ***Giants of Garbage.*** **Toronto: James Lorimer & Company. 300 pages.**

The names of the key players are not well known, but the men who developed companies such as Browning Ferris Industries, Laidlaw, and Waste Management, Inc., are part of a global corporate presence that Crooks accuses of collusion, unfair pricing, and monopolizing the waste trade. While the companies have paid millions of dollars in fines, Crooks writes that the waste industry routinely violates and skirts the law, is linked to organized crime, and has enormous economic and political power that results in little, if any, consequences.

Davis, Charles E. 1993. ***The Politics of Hazardous Waste.*** **Englewood Cliffs, NJ: Prentice-Hall. 156 pages.**

The public's fears about the dangers of hazardous waste make it difficult for policy makers to develop programs for dealing with it, even though many experts believe it is not the most serious environmental problem officials face. Davis provides an overview of the problem and how finding ways to deal with hazardous waste is complicated by emotion and controversy.

Duffy, John. 1990. ***The Sanitarians: A History of American Public Health.*** **Urbana: University of Illinois Press. 330 pages.**

The early beginnings of the U.S. Public Health Service and of sanitary engineering as a profession are not well known. This book outlines the development of concerns about the relationship between urban growth and health, and of the individuals who promoted the idea that a city could be cleaned up using both theory and technology.

El Haggar, Salah. 2007. ***Sustainable Industrial Design and Waste Management.*** **Amsterdam: Elsevier. 416 pages.**

Industry-generated waste is now a part of the development of corporate environmental responsibility. The author suggests that all waste can be treated from "the cradle to the grave" as a way of conducting sustainable business practices and reducing the impact on the depletion of natural resources.

Engler, Mira. 2004. ***Designing America's Waste Landscapes.*** **Baltimore, MD: Johns Hopkins University Press. 279 pages.**

The author begins by noting that landfills, more commonly called dumps, need to be viewed in terms of space, rather than as "lost" areas of little to no value. She shows that the areas can become a part of landscape design, transformed from areas of waste commerce to gardens and resource parks.

Ferrell, Jeff. 2006. ***Empire of Scrounge: Inside the Urban Underground of Dumpster Diving, Trash Picking, and Street Scavenging.*** **New York: New York University Press. 222 pages.**

Those who live on the margins of society must use every strategy possible to survive, Ferrell notes, including raiding dumpsters looking for scraps of food and salvaging usable items from trash cans outside buildings and homes. This view is written by a criminologist who provides an alternative, noncriminal view of street life.

Feshbach, Murray. 1995. ***Ecological Disaster: Cleaning Up the Hidden Legacy of the Soviet Regime.*** **New York: Twentieth Century Press. 157 pages.**

This book was written not too long after the records of the former Soviet Union began to be opened up and provides a snapshot of some of the massive problems facing this region. Hazardous waste from the Cold War and the military buildup within the country led to highly secretive programs to bury contamination, including high-level nuclear waste, throughout the Soviet Union. Feshbach provides insight into how serious the problem will be; even though the book is more than a decade old, still little is known about waste sites and disposal methods, especially in the Arctic region.

Fletcher, Thomas H. 2003. ***From Love Canal to Environmental Justice: The Politics of Hazardous Waste on the Canada-U.S. Border.*** **Peterborough, Ontario, Canada: Broadview Press. 239 pages.**

Using the frame of environmental justice, the author begins by examining a waste management crisis in Toronto that involved a proposal to ship municipal garbage by rail to a rural area in Ontario that was home to First Nations peoples who did not want the garbage dumped there. He also covers the infamous Love Canal incident and its role in the development of environmental policy, and Canada's efforts to avoid becoming a "waste haven" for other countries. Border regions such as Detroit and Niagara, New York, are used as case studies.

Fullerton, Don, and Thomas C. Kinnaman, eds. 2002. ***Economics of Household Garbage and Recycling Behavior.*** **North Hampston, MA: Edward Elgar. 224 pages.**

The editors of this book have compiled a series of studies to show what needs to be done to make recycling economically feasible and politically acceptable. The econometric techniques and theoretical solutions provided are a bit difficult for the average reader to understand, but the book also provides clear recommendations for municipal governments that are looking at new ways to make recycling a regular part of waste management in American households.

Garbutt, John. 2005. ***Waste Management Law: A Practical Handbook.*** **London: Sweet & Maxwell. 312 pages.**

The focus of this book is the United Kingdom, which provides a valuable comparison to U.S. and Canadian practices. The background information is primarily historical, providing the reader with a more extensive analysis than most reference books. This book is useful for those readers looking for a comparative approach to waste management and for updated material on a rapidly changing technology.

Gephardt, R. E. 2003. ***Hanford: A Conversation about Nuclear Waste and Cleanup.*** **Columbus, OH: Battelle Press. 384 pages.**

Washington State's Hanford site provides a case study in nuclear waste storage and site remediation. The author focuses on the environmental aspects of radioactive waste and its disposal in a neutral tone without obvious bias. While the book does not go into great depth in explaining the plight of the Downwinders and others who were exposed to nuclear production, it does provide an explanation of the key issues that policy makers need to address.

Girdner, Eddie J., and Jack Smith. 2002. ***Killing Me Softly: Toxic Waste, Corporate Profit, and the Struggle for Environmental Justice.*** **New York: Monthly Review Press. 162 pages.**

Mercer County, Missouri, is used as a case study to illustrate how the citizens of one community fought the efforts of Amoco Waste-Tech to build a hazardous waste landfill there between 1990 and 1993. The authors contend that the residents successfully opposed the project despite the fact that they were poor and politically marginalized because they were determined not to let their community become a "sacrifice zone" for the waste of the rest of the world.

Hamblin, Jacob Darwin. 2008. ***Poison in the Well: Radioactive Waste in the Oceans at the Dawn of the Nuclear Age.*** **New Brunswick, NJ: Rutgers University Press. 320 pages.**

The demise of the former Soviet Union has produced revelations about that country's environmental policies, including disclosures about decades of dumping radioactive waste into the ocean, rivers, and streams. This historical treatment of radioactive waste begins with the end of World War II and continues through the Cold War until the 1970s. It emphasizes the reasons why military strategy and political game playing allowed Soviet leaders to criticize the United States for its practices while at the same time covering up blatant abuses of international law.

Imhoff, Daniel. 2005. ***Paper or Plastic? Searching for Solutions to an Overpackaged World.*** **San Francisco: Sierra Club Books. 168 pages.**

The modern consumer choice between paper and plastic symbolizes what the author believes is the underlying question of sustainability. Do we make the decision based on cutting down trees or supporting the petrochemical industry? The statistics cited on the amount of municipal waste that is composed of packaging make clear that an equal amount of responsibility is to be shared among product designers, retailers, and consumers. The book presents numerous case studies of ways in which products can be packaged to reduce or eliminate waste and the role of packaged goods in society.

Jacobson, Timothy C. 1993. ***Waste Management: An American Corporate Success Story.*** **Washington, DC: Gateway Business Books. 340 pages.**

Most of the books that have been written about the refuse disposal industry are negative in their appraisal of the financial aspects of waste management. This book is different because it uses cases within the United States to show how companies have used their expertise to find new ways of disposing of waste to make practices more profitable.

Justice and Witness Ministries. 2007. ***Toxic Wastes and Race at Twenty: 1987–2007.*** **New York: United Church of Christ. 175 pages.**

The follow-up to the 1987 United Church of Christ study uses census data from 2000 and a database of commercial hazardous waste facilities to assess the extent of disparities in facility siting. The report found that people of color make up most of the residents living in host neighborhoods within 3 kilometers (1.8 miles) of the United States' facilities and that poverty rates are 1.5 times greater than in nonhost areas.

Kinnaman, Thomas C. 2003. ***The Economics of Residential Solid Waste Management.*** **Burlington, VT: Ashgate. 423 pages.**

This book is part of a series on environmental economics and policy, using refuse disposal within the United States as the case study. Although it provides a substantial amount of discussion about the advent of the refuse industry, it also deals with recycling and waste minimization and the impact of those practices on the industry.

Louka, Elli. 1994. ***Overcoming National Barriers to International Waste Trade: A New Perspective on the Transnational Movements of Hazardous and Radioactive Wastes.*** **Boston: Graham & Trotman. 226 pages.**

In developing an overview of a subject, it is often useful to look at older books that provide context for what is going on now. This book on hazardous waste identifies the early laws on international waste trade and explains how difficult it is to work with developing countries that view waste only in terms of dollars and cents.

Luton, Larry S. 1996. ***The Politics of Garbage.*** **Pittsburgh: University of Pittsburgh Press. 307 pages.**

Luton may be overly optimistic about the role of solid waste management, which he terms "one of the largest economic, ecological, and intellectual challenges faced in the United States today," as he tries to make a case why garbage should matter. He uses the case of Spokane, Washington, to show how the history and structure of local government affect decisions regarding solid waste management, and the role of other stakeholders and citizen participation.

Macey, Gregg P., and Jonathan Z. Cannon. 2007. ***Reclaiming the Land: Rethinking Superfund Institutions, Methods and Practices.*** **New York: Springer. 265 pages.**

This provocative book contends that while federal agencies are cleansing toxic waste sites of dangerous substances, they are also involved in what one researcher calls an unfortunate exercise of cultural and historical amnesia. Reclaiming brownfields, for instance, may provide opportunities for redevelopment, but at the cost of eliminating important landmarks where residents lived and have important memories of what took place.

Manser, A. G. R. 1996. ***Practical Handbook of Processing and Recycling Municipal Waste.*** **Boca Raton, FL: CRC/Lewis Publishers. 557 pages.**

This lengthy volume covers virtually every conceivable way that municipal waste can be handled. It discusses composting in the greatest depth, from simple window composting systems to mechanical composting and combined compost/recycling facilities. It also provides a rationale for the development of national standards legislation, although the ideas are somewhat dated.

Mason, Leonora G. 2007. ***Focus on Hazardous Materials Research.*** **Hauppauge, NY: Nova Science Publishers. 234 pages.**

This is one of a series of books from this publisher that highlights the studies and findings on the most recent issues in science. The other books are much more technical than this one, which has six chapters by international researchers from Spain, India, Italy, and Japan on technological advances in dealing with hazardous waste. The most accessible chapters are the ones dealing with risk assessment and the use of seaweed in phytoremediation.

McCutcheon, Chuck. 2002. ***Nuclear Reactions: The Politics of Opening a Radioactive Waste Disposal Site.*** **Albuquerque: University of New Mexico Press. 231 pages.**

Little material is available on the Waste Isolation Power Plant (WIPP) in southeastern New Mexico, in part because of the emphasis on controversies surrounding the Yucca Mountain facility in Nevada. The author uses a chronological approach in explaining how the WIPP was proposed, the opposition that arose, the decades spent in testing and study, and the eventual opening of the facility.

McDonough, William, and Michael Braungart. 2002. ***Cradle to Cradle: Remaking the Way We Make Things.*** **New York: Farrar, Strauss and Giroux. 208 pages.**

The title of this book comes from the concept of tracking materials from the cradle to the grave, meaning from the creation to the disposal of the substance. The authors contend that this approach still is wasteful and does not consider the option of designing products so that after their useful life is over, they can become biological nutrients or technical nutrients. Ideally, all materials could be used in perpetuity as part of an industrial re-revolution.

McFarlane, Allison, and Rodney C. Ewing, eds. 2006. ***Uncertainty Underground: Yucca Mountain and the Nation's High-Level Nuclear Waste.*** **Cambridge, MA: MIT Press. 431 pages.**

The 24 chapters in this edited volume are primarily science based, offering an analysis of the various issues under consideration for the operation of the Yucca Mountain nuclear waste repository in Nevada. Factors include the site's geologic stability and volcanism, hydrology, thermohydrology, waste packaging, and the difficulties of coping with uncertain models of operation.

Medina, Martin. 2007. ***The World's Scavengers: Salvaging for Sustainable Consumption and Production.*** **Lanham, MD: AltaMira Press. 318 pages.**

One of the most telling statistics in this book is that 2 percent of the world's population, an estimated 64 million people, survives by scavenging. The author uses case studies from Mexico, Brazil, Columbia, Argentina, Egypt, and the Philippines to show that scavenging is not a marginal economic activity relegated to the poor, and that it is an adaptive response to poverty. The coverage is both historical and contemporary, evidence that an informal economy is a part of the ingrained culture in many nations.

Melfort, Warren S. 2003. ***Nuclear Waste Disposal: Current Issues and Proposals.*** **Hauppauge, NY: Nova Science Publishers. 150 pages.**

Many supporters of nuclear power as a source of clean energy admit that the issue that poses the biggest problem is what to do with the waste that results. Since many U.S. nuclear facilities are at or near the end of their original licensing periods or operating life, the topic of how to decommission the plants and dispose of their nuclear components is becoming increasingly important. A second problem, the disposal of nuclear waste from weapons, is also discussed in this book, as are innovative cleanup technologies.

Melosi, Martin V. 2001. ***Effluent America: Cities, Industry, Energy, and the Environment.*** **Pittsburgh: University of Pittsburgh Press. 325 pages.**

Melosi attempts to cover an especially broad range of urban issues, building on past articles and book chapters he has written over nearly 25 years of research on the urban environment. Some of the material is slightly dated and repetitive, but the book's 12 chapters provide a good overview of the subject from a historical perspective.

Miller, Benjamin. 2000. ***Fat of the Land: The Garbage of New York—The Last Two Hundred Years.*** **New York: Four Walls Eight Windows. 414 pages.**

The author is the former head of the policy division of the New York Department of Sanitation and brings both a historical and a political eye to the topic of the city's trash. Although he begins with the stories behind the city's incineration program and the Fresh Kills landfill, he brings in the contemporary view of how municipal politics have influenced decision making and garbage services. Much depends, he notes, on how the problem of waste is framed as a policy problem, rather than technical solutions.

Mogren, Eric W. 2002. ***Warm Sands: Uranium Mill Tailings in the Atomic West.*** **Albuquerque: University of New Mexico Press. 241 pages.**

The atomic age has left an ugly legacy in the states of the Colorado Plateau—hundreds of acres of piles of uranium waste tailings that remain near the mines used to fuel nuclear plants. Most sites have now been cleaned up, but the author explains how the health and environmental aspects of mining waste continue to plague residents and mine workers.

Moore, Emmett B. 2000. ***An Introduction to the Management and Regulation of Hazardous Waste.*** **Columbus, OH: Battelle Press. 139 pages.**

The author has produced an excellent, short introduction that covers pertinent legislation, the chemistry of hazardous waste, an overview of various substances, pollution-control technologies and cleanup techniques, risk assessment, and other related statutes and regulations. He concludes with a section on liability for violations of federal environmental laws and the issues involved with siting hazardous waste facilities.

Murdock, Steven H., Richard S. Krannich, and F. Larry Leistritz. 1999. ***Hazardous Wastes in Rural America: Impacts, Implications, and Options for Rural Communities.*** **Lanham, MD: Rowman and Littlefield. 233 pages.**

Six chapters in this book represent the results of a multisite and multistage analysis of the impacts of hazardous waste siting. The authors collected data on 15 communities in five states to examine what factors (economic, demographic, public service, fiscal, and social and special) impacted the siting of facilities in the Great Plains and Intermountain West. Although they admit to the limitations of the sample, they conclude that substantial resistance is likely in every community when these facilities are proposed.

Myers, Garth Andrew. 2005. ***Disposable Cities: Garbage, Governance, and Sustainable Development in Urban Africa.*** **Aldershot, UK: Ashgate. 204 pages.**

This study of three cities—Dar es Salaam and Zanzibar, Tanzania and Lusaka, Zambia—provides an analysis of the United Nations' Sustainable Cities Programme, focusing on solid waste management. The author, from the University of Kansas, concludes that waste policies are influenced largely by the cultural and political histories of the urban landscape in the region.

National Research Council, Committee on Disposition of High-Level Radioactive Waste through Geological Isolation, Board on Radioactive Waste Management. 2001. ***Disposition of High-Level Waste and Spent Nuclear Fuel: The Continuing Societal and Technical Challenges.*** **Washington, DC: National Academies Press. 212 pages.**

This publication is a nontechnical summary of the issues facing government leaders in dealing with high-level radioactive waste. The major recommendation is that disposition in a deep geological repository is a sound approach as long as it progresses through a stepwise decision-making process that takes advantage of technical advances, public participation, and international cooperation.

National Research Council, Committee on Separations Technology and Transmutation Systems. 1996. ***Nuclear Wastes:***

Technologies for Separations and Transmutation. **Washington, DC: National Academies Press. 592 pages.**

This document was produced before the geologic depository at Yucca Mountain, Nevada, was chosen by the George H. W. Bush administration for the disposal of high-level nuclear waste, but it provides information on other options that were being considered prior to that decision. Its emphasis is on plutonium extraction for commercial reactor spent fuel in a time when the United States was concerned about issues of health, safety, and nuclear proliferation.

Nemerow, Nelson. 2006. ***Industrial Waste Treatment.*** **Oxford, UK: Elsevier. 568 pages.**

This lengthy book is targeted more to engineers and managers of industrial waste treatment facilities than to a general audience. It does, however, provide a thorough, if technical, overview of the changes that the U.S. government regulatory system is making on facilities. It begins by providing a summary of the history of environmental waste in the 20th century and practices such as volume reduction and neutralization. The second part of the book covers 21st century issues, such as zero waste, municipal-industrial complexes, and cost-benefit analyses for industrial waste facilities.

O'Brien, Martin. 2007. ***A Crisis of Waste? Understanding the Rubbish Society.*** **New York: Routledge. 210 pages.**

The author's perspective in this book is both historical and cultural. It attempts to determine why waste is spoken of in terms of crisis. He explains that what society considers to be waste is based in part on the level of industrialization that takes place. Kitchen grease, for instance, is used in many households as the basis for sauces or reused for lubrication, while in other homes it is thrown out as worthless.

Oh, Chang H. 2001. ***Hazardous and Radioactive Waste Treatment Technologies Handbook.*** **Boca Raton, FL: CRC Press. 792 pages.**

Despite the book's title, this volume has useful information for nontechnicians on the regulation of hazardous chemical and radioactive waste. Although it is primarily a reference book, it does provide an overview of recent waste technologies and remediation strategies, research, and development. It covers the role of the U.S. Department of Energy and the U.S. Department of Defense and the regulation of nuclear waste.

O'Neill, Kate. *Waste Trading among Nations: Building a New Theory of Environmental Regulation.* 2000. Cambridge, MA: MIT Press. 298 pages.

This award-winning book outlines the crisis in hazardous waste management—waste trading—more than 80 percent of which takes place among industrialized nations. Unlike the activities controlled by international regimes, which ban the export from developed countries to developing ones, this type of trade is, for the most part, entirely legal. The author focuses attention on the United Kingdom, Germany, France, Australia, and Japan.

Onibokun, Adepoju G., ed. 1999. *Managing the Monster: Urban Waste and Governance in Africa.* Ottawa, Canada: International Development Research Center. 270 pages.

The authors of the chapters in this volume show how growing urbanization in Africa has led to challenges in the ways officials manage waste. The conflict between the costs charged to users and the need for increasing urban revenues makes it difficult for the major cities that are used as case studies by the researchers to keep pace with waste deposits and disposal.

Page, G. William. 1997. *Contaminated Sites and Environmental Cleanup: International Approaches to Prevention, Remediation, and Reuse.* San Diego: Academic Press. 212 pages.

The title of the book implies that it covers a broader sample than it actually does, as only countries within the European Union are discussed. The major focus is on how countries approach sites contaminated with toxic waste, analyzing efforts at remediation from the public health, liability, and cleanup standpoints. A small section of the book is devoted to developing countries, and information is provided about Superfund cleanup process in the United States. The primary conclusion is that no international uniformity of policy exists toward toxic contamination and environmental cleanup, with some countries having no policy at all.

Pellow, David N. 2004. ***Garbage Wars: The Struggle for Environmental Justice in Chicago.*** **Cambridge, MA: MIT Press. 256 pages.**

The garbage industry makes for an interesting social, political, and historical story, especially when the setting is in Chicago. The author covers the period from 1880 to 2000, explaining how the health risks and environmental burden of waste fall disproportionately on communities of color and poverty. Also included are segments covering the efforts of residents to resist the siting of hazardous waste facilities, incinerators, and landfills.

Pellow, David N. 2007. ***Resisting Global Toxics: Transnational Movements for Environmental Justice.*** **Cambridge, MA: MIT Press. 336 pages.**

The topic of dumping wastes in developing countries has inspired a large number of recent books, all of which condemn the practice. This author explores the idea that hazardous waste dumping is becoming increasingly racialized and is now the focus of environmental justice groups that seek transnational solutions to reverse the practice. The emphasis on race and class is what distinguishes this book from others.

Peterson, Thomas V. 2002. ***Linked Arms: A Rural Community Resists Nuclear Waste.*** **Albany: State University of New York Press. 266 pages.**

This book on low-level nuclear waste facilities reads more like a novel than a scientific review, outlining the struggles of a community in New York that used civil disobedience in an attempt to stop the creation of a disposal facility there. It covers both the environmental protection arguments related to siting such a facility and the role of citizen participation in decision making, including the actions that led to community members' arrest.

Pezzullo, Phaedra C. 2007. ***Toxic Tourism: Rhetorics of Pollution, Travel, and Environmental Justice.*** **Tuscaloosa: University of Alabama Press. 265 pages.**

The author writes about the phenomenon of using tourism for politically progressive ends, contending that noncommercial tours can help mobilize public sentiment and dissent. The examples used are places where tours of toxic sites are provided by industries (also called plant tours), but more commonly, expeditions into places that are polluted by toxins. Local residents can educate outsiders about the proximity of homes and schools to industrial sites and can personalize poverty for those who do not see it on a regular basis. Toxic tourism has become an element of the environmental justice movement, focusing attention on accountability and public relations.

Piasecki, Bruce W., and Gary A. Davis, eds. 1987. ***America's Future in Toxic Waste Management: Lessons from Europe.*** **New York: Quorum. 325 pages.**

This edited volume has 10 essays, some of which compare the U.S. system of waste management with cross-cultural studies from Europe. The emphasis is on hazardous and toxic waste, although some references are made to waste reduction, technological innovation, and risk management.

Porter, C. Richard. 2002. ***The Economics of Waste.*** **Washington, DC: Resources for the Future. 240 pages.**

As a natural resource issue, the way in which economics affects waste management is not a topic many politicians would like to think about. The author contends that ways of dealing with both municipal and industrial waste are available that are influenced by market forces. The chapters cover issues such as hazardous waste trading in an international setting, recycling, the Superfund program, nuclear waste, and hazardous materials. The terminology is sometimes difficult to comprehend unless the reader has a background in economics.

Probst, Katherine N., and Thomas C. Beierle. 1999. ***The Evolution of Hazardous Waste Programs: Lessons from Eight Countries.*** **Washington, DC: Resources for the Future. 102 pages.**

This World Bank–funded study explains how four developed countries (Germany, Denmark, the United States, and Canada) and four developing countries (Malaysia, Hong Kong, Thailand, and Indonesia) have dealt with the problem of hazardous waste. This cross-country comparison shows the importance of clear rules and adequate enforcement and explains why companies have little incentive to comply with regulations without those factors. The authors conclude with one clear lesson: It takes a long time to develop an effective hazardous waste program.

Rahm, Dianne, ed. 2002. ***Toxic Waste and Environmental Policy in the 21st Century United States.*** **Jefferson, NC: McFarland & Co. 184 pages.**

The seven chapters in this short volume deal with a variety of toxic waste issues, from chemical and nuclear weapons disposal to the U.S. Environmental Protection Agency (EPA) and hazardous waste. The weakness of the book is that the topics are not tied together well, with some (such as community-based watershed remediation) seeming out of place.

Rathge, William, and Cullen Murphy. 1992. ***Rubbish! The Archaeology of Garbage.*** **New York: HarperCollins. 263 pages.**

The Garbage Project at the University of Arizona, which started in 1972, is the basis for this book, which provides a unique insight into the garbage collected in family trash cans as well as landfills as a way of chronicling American life. The book covers more than just the dumping of trash, however, explaining why some communities recycle while others do not, and what the students in the project found as they sifted through municipal dumps.

Riley, Peter. 2004. ***Nuclear Waste: Law, Policy, and Pragmatism.*** **Aldershot, UK: Ashgate. 300 pages.**

The author intends to provide a neutral framework for understanding the complexity of dealing with decades of nuclear waste. He focuses on government policies, both at the U.S. and international level, and explains the sources of laws and how they have been affected by stakeholders. The case studies deal with the United Kingdom, the United States, France, Finland, and Korea, with a heavy emphasis on court cases.

Ringius, Lasse. 2001. ***Radioactive Waste Disposal at Sea: Public Ideas, Transnational Policy Entrepreneurs, and Environmental Regimes.*** **Cambridge, MA: MIT Press. 261 pages.**

While dumping low-level radioactive waste in the oceans may seem like an idea whose time should never come, it is one option that has been seriously considered for decades. This book is useful because it not only explains the environmental aspects and science of marine disposal but also reviews the way in which international agreements (regimes) are developed. The author calls for international cooperation in changing global ocean dumping strategies.

Rogers, Heather. 2005. *Gone Tomorrow: The Hidden Life of Garbage.* New York: New Press. 288 pages.

The author's chronology of the history of garbage covers the development of the modern landfill, the concept of waste management and environmental harm, the successes and failures of recycling, and how garbage has become corporatized. The book was written as a follow-up to the author's 2002 documentary film, *Gone Tomorrow: The Hidden Life of Garbage,* and is written in a highly readable style for any audience interested in an overview of municipal waste.

Scanlan, John. 2005. *On Garbage.* London: Reaktion Books. 208 pages.

The philosophical side of trash is explored in this book, which looks at how humans view the artifacts they throw away. The author contends that Western culture has a unique perspective on what is "broken" in their lives, and how we see many of the objects we possess as useless debris. The writing style is quite different from most technical treatments of waste, but the book does provide insightful essays on culture.

Schwartz, Stephen C. 1983. *Energy and Resource Recovery from Waste.* Park Ridge, NJ: Noyes Data Corp. 272 pages.

The publication date of this book limits its applicability for contemporary discussion, but it does provide an excellent perspective on the refuse-to-fuel movement of the early 1980s. At that time, using garbage to produce energy was considered one of the most practical ways to recycle waste products, and the technology was available to make it a profitable enterprise. The author had no way of knowing that the idea of incinerating trash would become so controversial.

Smith, Paul G. ed. 2005. *Dictionary of Water and Waste Management,* 2nd ed. Boston: Butterworth-Heinemann. 480 pages.

This reference book was written, the author notes, for contractors, consultants, and professional engineers as well as academics and students. Although it is not especially technical, it is more suited for professional libraries than for daily use.

Smith, Ted, David A. Sonnenfeld, and David Naguib, eds. 2006. ***Challenging the Chip: Labor Rights and Environmental Justice in the Global Electronics Industry.*** **Philadelphia: Temple University Press. 376 pages.**

The impact of electronics manufacturing is global, the authors contend, with the negative consequences borne largely by the poor, women, immigrants, and minorities. The authors in this edited volume provide solutions for dealing with computer components and their waste, which contaminate and pollute the environment worldwide.

Stein, Kathy. 1997. ***Beyond Recycling.*** **Santa Fe, NM: Clear Light Publishers. 166 pages.**

This book may seem dated, but in fact, not much has happened to change the recommendations she makes since the book was published. This is a guide for those who want to locate companies that reuse products and others that have developed innovative ways of reducing waste. She explores the ways that consumers can find low-cost alternatives to disposables and how our choices to recycle can affect sustainability.

Strasser, Susan. 1999. ***Waste and Want: A Social History of Trash.*** **New York: Metropolitan Books. 368 pages.**

This is a social history of discarded items, from the colonial period to the time when scraps and rags were saved and sold, to mass production, marketing, and consumerism. Unlike more technical books that talk about the treatment of waste from a scientific or an engineering perspective, this one deals with the issues of waste and poverty, urban blight, ethnicity, and hoarding. The author also discusses the phenomenon of recycling and how it became emblematic of the counterculture but turned into a mainstream practice throughout society.

Tammemagi, Hans Y. 1999. ***The Waste Crisis: Landfills, Incinerators, and the Search for a Sustainable Future.*** **New York: Oxford University Press. 294 pages.**

Although one reviewer calls the topic of this book "unappealing," its historical coverage makes the book valuable for someone who needs an introduction to waste management. The discussion is limited to solid waste, focusing on the development of integrated waste management and recycling programs through seven case studies and proposed solutions. Perhaps the best chapter is the last one, which envisions the way garbage might be handled in the year 2032.

Tchobanoglous, George, and Frank Kreith, eds. 2002. ***Handbook of Solid Waste Management,*** **2nd ed. New York: McGraw-Hill. 950 pages.**

This massive volume is somewhat uneven because, while some of its 16 chapters are written in a readable style, most are so technical that only an engineer or a waste professional would understand the terminology and graphics. It does cover the entire spectrum of waste practices, from landfilling and source reduction to recycling and specialized waste such as scrap tires, batteries, and used oil. It includes an introductory section on federal and state waste management legislation and regulations.

Thompson, Ken. 2007. ***Compost: The Natural Way to Make Food for Your Garden.*** **New York: DK Publishing. 192 pages.**

The idea of starting a household compost system can be daunting, but this book offers practical ideas on how to deal with organic waste that can be used for the garden. The author writes for organic gardening magazines and shows how the best compost can be made out of ordinary household items.

U.S. General Accounting Office (GAO). 1983. ***Siting of Hazardous Waste Landfills and Their Correlation with Racial and Economic Status of Surrounding Communities.*** **Washington, DC: General Accounting Office.**

The issue of environmental justice was brought to the attention of policy makers with this report on four hazardous waste landfills in the U.S. South. The study showed that poor blacks comprised the majority of the population in all four communities where the facilities were sited. The GAO study generated protests at one of the sites in Warren County, North Carolina, in 1983.

Vallero, Daniel. 2003. ***Engineering the Risks of Hazardous Wastes.*** **Amsterdam: Elsevier. 306 pages.**

While numerous books are available on the operation of hazardous waste facilities, this is one of the few that discusses the risks of managing the wastes themselves. The focus is primarily on chemical wastes, but sections are also provided on dealing with employee perceptions of waste, emergency responses, and site remediation.

Vandenbosch, Robert, and Susanne E. Vandenbosch. 2007. ***Nuclear Waste Stalemate: Political and Scientific Controversies.*** **Salt Lake City: University of Utah Press. 464 pages.**

Advocates for the increased use of nuclear power as part of the goal of reducing U.S. reliance on imported fossil fuels will find this book valuable because of its emphasis on technological advances. It includes an excellent discussion of the political debate over nuclear waste as it applies to Yucca Mountain, the use of Native American reservations as host sites for spent nuclear fuel, and the use of incentives for accepting nuclear waste. Also provided is a sizeable section on long-term storage problems, developments in on-site versus geological storage, and the problems associated with seismic events.

Walsh, Edward J., Rex Warland, and D. Clayton Smith. 1997. ***Don't Burn It Here: Grassroots Challenges to Trash Incinerators.*** **University Park: Pennsylvania State University Press. 292 pages.**

The authors seek a sociological explanation for the grassroots anti-incineration movement that developed in the 1970s and 1980s as communities sought new ways of dealing with municipal waste. Trash incineration and waste-to-energy facilities were proposed as a way of saving landfill space, but many were never built as a result of local protests.

Weingart, John. 2007. ***Waste Is a Terrible Thing to Mind: Risk, Radiation, and Distrust of Government.*** **New Brunswick, NJ: Rutgers University Press. 448 pages.**

Although the title of this book is sarcastic, the author deals with deadly serious topics: radioactive waste and disposal. He looks at a variety of siting controversies and the role of local community groups and officials, some of whom have considered a voluntary siting process that involves regional and state-level agreements.

Westra, Laura, and Bill E. Lawson, eds. 2001. ***Faces of Environmental Racism: Confronting Issues of Global Justice.*** **2nd ed. Lanham, MD: Rowman and Littlefield. 266 pages.**

The 12 chapters in this edited book provide an overview of environmental justice issues both in the United States and globally. The book emphasizes inequities in environmental and health threats, which various authors contend is a result of policies that result in regressive patterns of costs and benefits.

Zimring, Carl A. 2006. ***Cash for Your Trash: Scrap Recycling in America.*** **New Brunswick, NJ: Rutgers University Press. 224 pages.**

While most people are familiar with residential recycling and curbside bins for their trash, industrial recycling predates contemporary recycling campaigns by decades. The author provides an analysis of the historical and practical development of recycling of items such as tires and rubber, metal, and glass.

Articles

Atlas, Mark. 2002. "Few and Far Between? An Environmental Equity Analysis of the Geographic Distribution of Hazardous Waste Generation." ***Social Science Quarterly*** **83 (1): 365–378.**

The research on environmental justice has traditionally focused on commercial facilities located in poor and minority neighborhoods, and this author contends that this focus ignores cases where facilities generate and manage their own waste. He believes that these operations, which account for 98 percent of the volume of hazardous waste, do not fit that pattern when population demographics are aggregated in concentric geographic rings around generating facilities.

Bae, Young-Ja. 2005. "Environmental Security in East Asia: The Case of Radioactive Waste Management." ***Asian Perspective*** **29 (2): 73–97.**

Economic growth in East Asia has been accompanied by increases in energy demand and an expansion of nuclear energy, which accounts for nearly a third of the production in Japan, Korea, and Taiwan. This article ties nuclear power to issues of environmental security, identifying the tensions that have developed over incidents such as nuclear waste dumping in the Sea of Japan and Taiwan's efforts to export nuclear waste to North Korea. Environmental security is, the author notes, a question of regional approaches where the prospects for cooperation must be addressed.

Biggert, Judy. 2006. "Nuclear Waste Standoff." ***Issues in Science and Technology*** **23 (1).**

In the Forum section of this issue, the author provides an overview of the Global Nuclear Energy Partnership (GNEP) and the issues of restarting the U.S. nuclear energy industry. The key problem remains how nuclear waste can be dealt with on both an interim and a long-term basis.

Black, George. 2007. "Waste Not, Want Not: American Hospitals Are Charter Members of Our Throwaway Society." ***OnEarth*** **(Winter): 26–32.**

An estimated 2,000 tons of medical equipment and supplies in perfectly good condition are tossed into the trash each year by U.S. hospitals. Hazardous waste goes to special incinerators, but the remainder goes to landfills, creating a specialized environmental problem. One organization, Global Links, began redistributing medical surplus and diverting it from the waste stream in 1990. Participating hospitals in Pennsylvania and West Virginia transfer supplies before they go to the dump, taking them to a warehouse for sorting before they are shipped to countries where the equipment and supplies are badly needed.

Clapp, Jennifer. 2002. "Seeping through the Regulatory Cracks." ***SAIS Review*** **21 (1): 141–155.**

An expert on hazardous waste trading, Clapp provides a history of the international trade in hazardous waste that began in the late 1980s up through the Basel Convention. She explains that key stakeholders are not parties to the convention (including the United States) and that this situation allows rich-country to poor-country exports of waste for recycling to continue. The lack of participation by industrialized nations is coupled with poor enforcement and the limited scope of the Basel agreement, with countries taking advantage of loopholes to export waste to developing countries.

CQ Researcher. 2006. "Momentum Stalls." _CQ Researcher_ 16 (10): 229–230.

This short piece explains why the momentum to build nuclear power plants in the United States seemed to stall despite the 1973 Arab oil crisis. This historical article covers the enactment of environmental laws, the growth of the environmental movement, waste management concerns, safety, and controversial incidents that stalled the momentum of nuclear power plant development.

Davidson, Pamela. 2000. "Demographics of Dumping II: A National Environmental Equity Survey and the Distribution of Hazardous Materials Handlers." _Demography_ 37 (4): 461–466.

The author examined the facilities governed by the Resource Conservation and Recovery Act (RCRA) throughout the United States to determine whether patterns occur in how these facilities are sited. The study found that although no evidence exists of stark environmental inequities, RCRA-governed facilities are more likely to be sited in working-class neighborhoods with lower percentages of minority residents but are also sited close to neighborhoods with a higher percentage of minority residents.

Friess, Steve. 2006. "A New 'Joe Camel'?" _Newsweek_ 147 (April 24): 9.

Officials in Nevada are outraged that the U.S. Department of Energy has created a cartoon figure to raise support for the storage of nuclear waste at the Yucca Mountain nuclear waste facility. They compare the attempt to promote the agency's project with the tobacco industry's use of the Joe Camel character to get children to smoke. Nevada's political leaders, who have overwhelmingly opposed the Yucca Mountain project, are venting their fury on the department.

Goldberg, Jessica. 2007. "The Nuke Waste Around Us." _E: The Environmental Magazine_ 18 (5): 24.

Radioactive contamination is now being found in U.S. landfills, according to a report by the Nuclear Information and Resource

Service. This article notes that radioactive materials from nuclear weapons production is ending up in municipal landfills, either directly released by the U.S. Department of Energy or by brokers and processors of the waste. The public is generally unaware of this problem, according to the author, and needs to become educated in order to tell the federal government to shift from the dispersal to the isolation of radioactive waste.

Hadjilambrinos, Constantine. 2006. "The High-Level Radioactive Waste Policy Dilemma: Prospects for a Realistic Management Policy." ***Journal of Technology Studies*** **32 (2): 95–105.**

Tremendous increases in the amount of high-level radioactive waste being generated in the United States require a coherent national policy, the author contends. The growth of nuclear power plants has resulted in waste that has the capacity to diminish the human species. Long-term containment policies have not yet resolved technical issues that still plague the nation.

Hickman, H. Lanier, and Richard W. Eldredge. n.d. "A Brief History of Solid Waste Management in the US, 1950–2000." ***Municipal Solid Waste Management.*** **[Online article; retrieved 7/1/07.] http://www.mswmanagement.com/msw.html.**

A series of articles that attempts to bring together the results of interviews with industry "elders," these recollections provide background material for a field that lacks many historical documents. It begins with a brief summary of the eradication of communicable diseases by the U.S. Public Health Service and moves through the development of sanitary landfills, the waste management infrastructure, waste-to-energy facilities, and the current status of solid waste as a profession and science.

Holton, W. Conrad. 2005. "Power Surge: Renewed Interest in Nuclear Energy." ***Environmental Health Perspectives*** **113 (11): 742–750.**

Although the nuclear power industry appears to have stalled for the last three decades, interest in developing nuclear plants seems to be renewed because of worldwide demands for energy. The U.S. Energy Policy Act, initiated by the George W. Bush administration, supports more research and development of nuclear power, and some environmental groups are taking a second look at this energy source. Despite potential benefits in slowing global climate change, nuclear power advocates must still come up with solutions to deal with the problems associated with nuclear waste.

Jenkins, R., Kelly B. Maguire, and Cynthia L. Morgan. 2004. "Host Community Compensation and Municipal Solid Waste Landfills." ***Land Economics*** **80 (4): 513–528.**

One of the newer incentives for convincing communities to accept landfills is payment by developers in exchange for permission to construct, expand, or operate facilities. This study finds that citizen participation in host fee negotiations, experience in hosting a landfill, state requirements for minimum hosting compensation, and firms with greater resources lead to great host compensation. Somewhat surprisingly, neither the racial makeup nor the income level of the community appears to be a factor in host siting decisions.

Kaufman, Scott, Nora Goldstein, Karsten Millrath, and Nickolas J. Themelis. 2004. "The State of Garbage in America." ***BioCycle*** **45 (1): 31–41.**

One of the difficulties encountered in solid waste management policy making is getting an accurate accounting of how much waste is actually generated in the United States. These authors examine two of the most frequently cited studies: the *BioCycle* State of Garbage in America survey and research conducted by Franklin Associates for the U.S. EPA. In order to deal with differences in the findings of the two studies, researchers collaborated to create a report on waste management from 1989 to 2002.

Kelly, Timothy M. 2005. "Can Hazardous Waste Sites be Breached as a Result of Future Climate Change?" ***Journal of Environmental Engineering*** **131 (5): 810–814.**

Among the many problems attributed to potential climate change is whether increasing precipitation will lead to serious consequences for the integrity of hazardous waste disposal facilities. This study shows that large-scale changes in rainfall have occurred over the Northern Hemisphere, and landfills have the potential for failure because they were designed for less precipitation. The problem may also apply to abandoned or closed hazardous waste sites, where little research has been conducted on the potential for breaching.

Kloosterman, Karin. 2007. "Israeli Discovery Converts Waste into Clean Energy." ***Israel21c*** **(March 18). [Online article; retrieved 8/12/07.] http://www.israel21c.org/bin/en.jsp?enZone=Technology&enDisplay=view&enPage=BlankPage&enDispWhat=object&enDispWho=Articles%5El1586.**

Israeli and Russian scientists have been working for years to find a new, cost-effective way to deal with radioactive and hazardous waste, and this article contends that they have developed a potential new way of reducing tons of waste to energy and glass. The process uses plasma gas melting technology to convert waste by treating it in a noncombustion manner that actually produces excess electric power.

Louis, Garrick E. 2004. "A Historical Context of Municipal Solid Waste Management in the United States." ***Waste Management and Research*** **22 (4): 306–322.**

This historical overview begins by noting that American cities lacked systems for street cleaning, refuse collection, waste treatment, and human waste removal until the early 1800s. The developers of the public works infrastructure did not turn their attention to solid waste until the 1880s, when refuse management was deemed a municipal responsibility, where it remains today.

Macauley, Molly, Karen Palmer, Jhih-Shyang Shih, Sarah Cline, and Heather Holsinger. 2001. "The Environment and the Information Age: The Costs of Coping with Used Computer Monitors." ***Resources*** **(Fall): 6–9.**

The authors identify the issues raised by the "information economy" and the growing use of personal computers. Although this article is now somewhat outdated, it does do a good job of analyzing disposal practices and the environmental concerns raised by computer monitor displays. It also recommends what might be done to develop electronic waste disposal policies based on various goals such as eliminating negative health effects and the costs of encouraging recycling.

Marxsen, Craig S. 2001. "Potential World Garbage and Waste Carbon Sequestration." ***Environmental Science and Policy*** **4 (December): 293–300.**

The emission of fossil fuels is just one way in which carbon dioxide is released into the atmosphere, exacerbating the problems of global climate change. This author contends that strategies are available for dealing with waste carbon, including that portion produced by garbage dumps, which involve sequestering it.

Melosi, Martin V. 2002. "The Fresno Sanitary Landfill in American Cultural Context." ***The Public Historian*** **24 (Summer): 17–35.**

Melosi, one of the United States' top garbage historians, explains the controversies surrounding the designation of the Fresno Sanitary Landfill as a National Historic Landmark on August 27, 2001. At issue was the fact that Secretary of the Interior Gale Norton rescinded the designation the following day because it was also a Superfund site. To some, the listing of the landfill was a national joke, while to others, it marked an appropriate way of honoring public works engineering.

Murphy, J. D., and E. McKeogh. 2004. "Technical, Economic and Environmental Analysis of Energy Production from Municipal Solid Waste." ***Renewable Energy*** **29 (7): 1043–1057.**

This study examines the production of energy from four technologies used to dispose of municipal solid waste: incineration, gasification, generation of biogas used to generate heat and power, and generation of biogas and conversion to transport fuel. The researchers conclude that biogas technologies are less expensive to operate than other methods and require lower initial investment fees.

Natural Resources Defense Council. 2006. "Toxic Pentagon." ***Earth Island Journal*** **20 (Winter): 8.**

The U.S. Department of Defense has been under fire for decades for its environmental policies, especially the cleanup of toxic and hazardous wastes. The Government Accountability Office has reported that the agency has failed to grant requests from both California and Texas to test sites for perchlorate contamination resulting from military activities. The chemical waste is especially dangerous to infants and is found in an estimated 35 states at 395 defense sites.

Nussbaum, Neil J. 2006. "Radiation Risk to Future Generations from Long-Lived Radioactive Waste." ***Journal of Community Health*** **31 (5): 363–368.**

This article provides an overview of the growth of the U.S. inventory of nuclear waste and the revision of storage standards by the U.S. Environmental Protection Agency. The delays in developing a permanent storage facility at Yucca Mountain, Nevada, present a significant radiation risk for future generations 10,000 years into the future.

Olivares, Cristina. 2007. "Laying Groundwork for Campus Composting." ***BioCycle*** **48 (June): 35.**

Ohio University students from a sustainable agriculture class embarked on a project to conduct a waste audit at the university as part of an effort to implement a full-scale composting program there. The university created the Office of Resource Conservation in 2006 to help organize the project. The initial findings show that the school will be able to divert 2.5 to 3 tons of food waste to composting each day on a campus with a population of about 17,000.

Peretz, Jean H., Bruce E. Tonn, and David H. Folz. 2005. "Explaining the Performance of Mature Municipal Solid Waste Recycling Programs." ***Journal of Environmental Planning and Management*** **48 (5): 627–650.**

Recycling programs began to be implemented in cities during the 1970s and 1980s, and this study evaluates how well they have worked. The researchers found that recycling participation rates were higher in cities that offered more convenient recycling programs and whose residents had a higher mean household income. Cities with a larger nonminority population that imposed sanctions on improper sorting of recyclable materials had a higher participation rate, and in small cities, higher participation was attained by mandating household participation.

Roach, John. 2003. "Are Plastic Grocery Bags Sacking the Environment?" ***National Geographic News*** **(September 2). [Online**

article; retrieved 4/26/07.] http://news.nationalgeographic.com/news/2003/09/0902_030902_plasticbags.html.

This update on the use of plastic bags explains why they have become such an environmental problem around the world. The author identifies the pros and cons of using plastic bags, including the fact that they are more economical. But he also notes how the bags end up in the marine environment where the debris is ingested by wildlife. The article concludes with the plastic bag levy adopted by the Republic of Ireland and whether it might be adapted to the United States.

Schultz, Max. 2006. "Power Is the Future." ***Wilson Quarterly*** **30 (4): 59–64.**

Advantages and disadvantages accompany the use of nuclear power, with critics saying that it is dangerous, and advocates, including environmental leader Patrick Moore, contending that it is the only large-scale, cost-effective energy source that can be implemented while the world looks at other ways of reducing greenhouse gas emissions. Global power demands are at odds with the problem posed by nuclear waste and the potential for the proliferation of nuclear material.

Slack, R. J., J. Gronow, and N. Voulvoulis. 2004. "Hazardous Components of Household Waste." ***Critical Reviews of Environmental Science and Technology*** **34 (5): 419–445.**

Hazardous waste is being found in increasing proportions of municipal solid waste, but little research is being conducted on methods used to dispose of this component of the waste stream. Because different countries classify household hazardous wastes differently, it is difficult to quantify how much is sent to landfills and the extent of the potential risks these wastes pose to both human health and the environment.

Wagner, Travis. 2004. "Hazardous Waste: Evolution of a National Environmental Problem." ***Journal of Policy History*** **16 (4): 306–331.**

The author explains that although hazardous waste has always been perceived as a major environmental problem and public health risk, no one took it seriously until 1978. Despite the U.S.

military's controversial program to dispose of outdated nerve gas weapons in the Atlantic Ocean in August 1970, called Operation CHASE, the public perceived the dumping as an isolated incident that really did not affect their backyards. This lack of awareness was one of the reasons why hazardous waste did not become a serious issue until Love Canal was widely covered by the media.

Journals

CIWM Journal
Chartered Institution of Wastes Management
9 Saxon Court
St. Peter's Gardens, Marefair
Northampton NN1 1SX
United Kingdom
Web site: www.ciwm.co.uk

EM
One Gateway Center, Third Floor
420 Fort Duquesne Boulevard
Pittsburgh, PA 15222-1435
Telephone: (412) 232-3444
Web site: www.awma.org

International Journal of Environment and Waste Management
Inderscience Enterprises
World Trade Center Building
29 Route de Pre-Bois
Case Postale 896
CH-1215 Geneva 15
Switzerland
Web site: www.inderscience.com

Journal of the Air & Waste Management Association
One Gateway Center, Third Floor
420 Fort Duquesne Boulevard
Pittsburgh, PA 15222-1435
Telephone: (412) 232-3444
Web site: www.awma.org

Journal of Packaging
Institute of Packaging Professionals
1601 Bond Street, Suite 101
Naperville, IL 60563
Telephone: (630) 544-5050
Web site: www.iopp.org

Journal of Solid Waste Technology and Management
Department of Civil Engineering
Widener University
One University Place
Chester, PA 19013-5792
Telephone: (610) 499-4042
Web site: http://www2.widener.edu/~sxw0004/solid_waste.html

MSW Management
Forester Communications, Inc.
P.O. Box 3100
Santa Barbara, CA 93130
Telephone: (805) 682-1300
Web site: http://www.mswmanagement.com/msw.html

Tribal Waste Management
U.S. Environmental Protection Agency
Municipal and Industrial Solid Waste Division (5306P)
Ariel Rios Building
1200 Pennsylvania Avenue NW
Washington, DC 20460
Telephone: (866) EPA-WEST or (415) 947-8000
Web site: www.epa.gov/tribalmsw

Waste and Resource Management
Institution of Civil Engineers
Thomas Telford Journals
One Great George Street
London SW1P 3AA
United Kingdom
Web site: www.thomastelford.com/journals

Waste Management and Research
Sage Publications

1 Oliver's Road, 55 City Road
London EC1Y 1SP
United Kingdom
Web site: www.iswa.org

Waste Management World
International Solid Waste Association
Vesterbrogade 74, Third Floor
1620 Copenhagen V
Denmark
Web site: www.iswa.org

Nonprint Resources

CD-ROMS/DVD-ROMS/DVDs/Videos

Approaches to Onsite Management: Community Perspectives
Type: Videorecording
Length: 17 minutes
Date: 2002
Source: National Environmental Services Center, West Virginia University, Morgantown, WV

Narrated by Harlan Hogan, this documentary video provides an introductory overview of the concept of community on-site/decentralized waste management and sewage disposal under the auspices of the Environmental Protection Agency's Office of Wastewater Management.

Borderline Cases: Environmental Matters at the U.S.–Mexico Border
Type: Videorecording
Length: 65 minutes
Date: 1997
Source: Bullfrog Films, Oley, PA

The growth of factories, known as maquiladoras, along the U.S. border with Mexico has created problems when solid and hazardous wastes are dumped by companies that fail to comply with environmental regulations.

Crapshoot: The Gamble with Our Wastes
Type: Videorecording
Length: 52 minutes
Date: 2004
Source: Bullfrog Films, Oley, PA

This global tour of waste and sewage sites provides reasons why the world needs to rethink its perspective on waste management.

Laid to Waste
Type: Videorecording
Length: 52 minutes
Date: 1996
Source: University of California Extension Center for Media, Berkeley, CA

One of the most important battles in the history of U.S. waste management took place in Chester, Pennsylvania, when community leaders and activists sought to stop the siting of a waste treatment facility in their city.

Radioactive America
Type: Videorecording
Length: 29 minutes
Date: 2000
Source: Center for Defense Information, Washington, DC

The United States is still dealing with the problems of leftover nuclear waste from the research conducted at the Oak Ridge National Laboratory in Tennessee and the Hanford site in Washington State.

Radioactive Reservations
Type: Videorecording or DVD
Length: 52 minutes
Date: 1995
Source: Filmmakers Library, New York, NY

Native Americans must choose whether they want to try to escape impoverished living conditions by agreeing to have their lands serve as the repositories for radioactive waste, even though tribal councils, rather than tribal members, may profit. Four reservations are profiled in negotiations with the U.S. government.

Ships of Shame
Type: Videorecording or DVD
Length: 28 minutes
Date: 2001
Source: Filmakers Library, New York, NY

In India, shipbreaking—the scrapping of older vessels—exposes poor laborers to environmental hazards and accidents, as seen in this documentary produced for the Association for Asian Studies. Thousands of exploited workers risk their lives to deal with the hazardous and dangerous waste of industrialized nations in what investigators believe is a case of international injustice.

The Ash Barge Odyssey
Type: Videorecording
Length: 57 minutes
Date: 2001
Source: Michael Thomas Productions, Montgomeryville, PA

In 1986, the world was captivated by the story of a barge loaded with 14,000 tons of incinerated trash that left Philadelphia looking for a landfill that would accept the waste. Although most of it was apparently dumped at sea, some was dumped in Haiti before the barge eventually returned to the United States.

The Trash Trade: Selling Garbage to China
Type: Videorecording or DVD
Length: 49 minutes
Date: 2004
Source: Filmakers Library, New York, NY

Japan sends its trash to China, where scrap dealers disassemble it and make it into new goods that are then shipped back to Japan in an international recycling program that contributes to Chinese pollution.

Toxic Waters
Type: Videorecording
Length: 58 minutes
Date: 2000
Source: Michael Thomas Productions, Montgomeryville, PA

Delaware County, Pennsylvania, is the unlikely venue for protest when citizens begin to rally around the issue of toxic waste in their neighborhood.

Two Square Miles
Type: Videorecording or DVD
Length: 52 minutes
Date: 2006
Source: Filmakers Library, New York, NY

The citizens of Hudson, New York, are profiled in this award-winning documentary that explores how a community responds to a proposed multinational coal-fired cement plant that threatens to generate environmental waste in a small town.

Glossary

aerobic digestion A process for breaking down organic material using bacteria to reduce the volume of municipal solid waste, producing water and carbon dioxide.

anaerobic digestion The process used in landfills to reduce the volume of organic waste by serving as a medium for the growth in the absence of oxygen, converting the waste to methane and carbon dioxide.

ash The noncombustible material that remains after incineration, composed of fine powders, small parts of partially burned materials, and burned matter.

back-end process A technological action that changes the chemical properties of waste or converts its components into energy or compost, such as a waste-to-energy facility that converts biomass into electricity.

baler A device used in processing municipal solid waste or at similar facilities that compresses loose material into rectangular blocks. Retailers sometimes use balers to process cardboard boxes to make the material easier to handle and to avoid taking up space in trash receptacles.

biodegradable A product or substance that can be broken down or decomposed into basic organic materials, such as gases and acids.

biomedical waste Pathological or infectious waste, such as animal carcasses, bacteria and viruses, blood, or tissue.

bioremediation The use of natural microbial treatments (such as plants or microorganisms) to remediate toxic waste contamination.

blunt policy instruments Policies that indirectly influence potentially polluting activities, such as the dumping of waste, by establishing a tax on inputs that, after some industrial activity, will generate toxic wastes.

brownfields Sites for potential redevelopment that have been previously used but are not now in productive use, often within urban areas.

bypassed waste Waste received at a waste-to-energy facility that is not burned because it is noncombustible, unacceptable, or too large, such as automobile batteries or discarded refrigerators.

clinker Material that is formed as a result of a fire that is usually hard and fused together, such as glass or metal, that cannot be completely burned.

cogeneration The process by which the heat released from one operation serves as a source of energy for another operation, such as the steam that is produced in a boiler that is captured and used to power an electrical generator.

combustion Combining oxygen with a substance to produce heat, such as fire or light.

command and control The most common means of regulating the use and disposal of hazardous materials and wastes. This often includes policies and restrictions that specify a maximum allowable level of pollution or toxicity.

compaction garbage collection The use of a collection truck that was designed to hydraulically compress household waste inside the closed portion of the vehicle, requiring fewer workers and allowing a vehicle to collect more trash in each trip before unloading at the landfill.

composting The controlled biological decomposition of waste. Compost is made of materials such as food; leaves, grass, and other yard waste; paper; and cardboard. If produced under controlled conditions, it can be used to enrich garden soil.

contaminated sites Parcels of land on which or under which hazardous and toxic substances exist under conditions that do not effectively confine their movement.

cradle to grave A policy approach that uses manifests to control hazardous wastes from the point of generation to the final disposal sites. The controls apply to hazardous waste generators and transporters as well as those that store, treat, or dispose of the wastes.

cullet Pieces of used glass broken up for recycling that are sometimes melted to make new glass products or combined with other products.

deep pockets In litigation over which parties are legally responsible for hazardous waste cleanup, the one or more litigants who have the greater capacity to pay for the costs involved.

de-inking The process of removing ink, filler, and other extraneous material from printed or unprinted recovered paper to produce pulp, which can be used in the manufacture of new paper or tissue.

densification A process that lowers the volume-to-weight ratio of waste in order to reduce shipping costs. Baling is the most common form of densification, although some handlers of postconsumer plastics grind materials up for recycling.

dump A generic name for the indiscriminate depositing of waste materials without monitoring or regulation. Dumps are usually open sites where waste is buried or left to decompose.

extraction treatments The physical removal of contaminated soil, surface water, or groundwater from a site and then confining or treating it.

fly ash Finely divided carbon particles, cinders, dust, soot, or other partially burned materials that rise with hot combustion gases and are trapped at the back end of waste-to-energy plants.

front-end recovery Methods used to reduce the size of solid waste by separating or physically modifying components so they can be used or reused, such as separating glass bottles before they are landfilled.

handler A person or group that prepares recyclable plastics or other materials by sorting, densifying, or storing the material until a sufficient quantity is on hand.

hauler A person or company that delivers trash and postconsumer materials to a transfer station or processing facility, such as a landfill or an incinerator. Waste haulers may be public or private entities that contract to the public to provide trash services.

hazardous Describing a substance, such as waste, that has one or more of the following characteristics: toxicity, corrosivity, flammability, or reactivity.

incineration A form of waste disposal that involves burning refuse at extremely high temperatures for treatment.

integrated solid waste management The use of a combination of waste management techniques that ranks options in order of their preference as ways of dealing with waste. The common preference order is source reduction, recycling and composting, combustion, and landfilling.

joint and several liability A component of hazardous waste cleanup responsibility whereby any one of the potentially responsible parties can be liable for the entire cleanup cost at a site. This concept allows the government to collect the costs of cleanup with a powerful tool when it is not possible to find all of the parties.

landfill A large land area designed, operated, and maintained for depositing solid waste. Landfills in the United States are regulated by the Environmental Protection Agency.

landfill gas The emissions of decomposing waste that occur in landfill operations, usually consisting of highly flammable methane, a major contributor to global climate change. It may also include the airborne wastes from adhesives, household cleaners, plastics, and paints.

leachate Types of liquids that pass through solid waste, picking up dissolved or suspended particles. Landfills trap leachate by using a liner as a barrier to minimize the possibility that it will mix with ground or surface water.

materials recovery The separation of resources such as paper, glass, and metals with the purpose of processing the waste for beneficial use or reuse. Materials recovery facilities are plants that use this strategy to separate and sort municipal solid waste.

mega-fills Underground waste sites that range in size from 10 to 100 acres across and hundreds of feet deep.

methane An odorless, colorless, nonpoisonous, flammable, and explosive gas formed when organic waste decomposes. Methane is often produced at landfills, where it can be recovered and used as fuel to avoid having it enter the atmosphere.

midnight dumping Illegal disposal of hazardous or other types of waste in nonapproved disposal sites. This practice often occurred before the government adopted the cradle-to-grave policy to track the final disposition of wastes, and included trucks that would spray hazardous waste oil on country roads or dump waste in a pit or river in the dark of the night to avoid detection.

mine-scarred lands Areas and their associated waters and surrounding watersheds where extraction, beneficiation, or processing of ores and minerals has occurred. These properties may be on public or private land and involve complex environmental and social issues. Some mine-scarred lands may be reused as landfills.

municipal solid waste Any household waste or garbage, street litter, or solid waste produced by businesses, institutions, and schools. It does not include materials separated out for recycling, infectious or hazardous waste, waste from industrial processes, demolition or disaster debris, mining or agricultural waste, ashes, or sewage sludge.

National Priorities List A list of the most dangerous hazardous waste sites in the United States that will receive an environmental cleanup under the Superfund program.

open burning The combustion of solid waste without controls on air emissions. This is often a method for waste disposal in rural areas where trash collection is unavailable or expensive.

orphan site Commercial land contaminated by hazardous waste for which no responsible party can be identified or is capable of remediating the site.

packaging The wrapping, covering, or container used to store, carry, or display a product. The majority of solid waste comes from packaging, especially paper and plastic wrapping.

percolation The movement of water downward through solid waste into the soil below. Landfills use liners and drainage pipes to prevent percolation.

petroleum brownfields Areas where underground storage tanks are found on abandoned or underused industrial and commercial properties. These plots are being cleaned up with funds from the Leaking Underground Storage Tank Trust Fund for reuse.

phytoremediation The direct use of green plants and their associated microorganisms to stabilize or reduce contamination in soils, sludges, sediments, surface water, or ground water. This process was first used in the 1990s to clean up hazardous waste sites.

polluter-pays principle The concept that the party responsible for causing the pollution should bear the cost of efforts to reduce or remediate the pollution, used as a way of distributing the costs of environmental cleanup or protection.

portfields Brownfields (contaminated industrial areas) near port communities. Cleanup of these waste sites in the United States is led by the National Oceanic and Atmospheric Administration and the Environmental Protection Agency, with support from the Army Corps of Engineers.

potentially responsible parties Term used under the U.S. Comprehensive Environmental Response, Compensation, and Liability Act to determine who may be responsible for the cost of cleanup of a hazardous waste site.

proximity principle The concept that wastes should be transferred or transported to the nearest, best disposal facility, regardless of jurisdiction.

pyrolysis An industrial process involving the destructive distillation of a solid material using heat, and, in the absence of oxygen, sometimes used for solid waste disposal.

railfields Lands that have been abandoned with the decline of the railroad industry during the 20th century that have been contaminated with waste, especially hazardous waste. Railfields in the United States are being cleaned up and redeveloped by the Environmental Protection Agency.

remediation The process of environmental cleanup on a contaminated site and the techniques to decrease or eliminate contamination from soil, surface water, or groundwater.

resource recovery The burning of wastes to create steam power and electricity from combustion using materials that would otherwise be discarded. Resource recovery facilities are also called waste-to-energy plants.

retroactive liability A legal principle used under the Superfund program that makes parties liable to pay the cost of environmental cleanup for acts that were legal at the time they occurred but that caused contamination.

sanitarians Professionals within the refuse collection and disposal management systems during the late 19th and early 20th centuries.

sanitary landfill An engineered facility for the disposal of municipal solid waste where efforts are made to minimize the negative environmental and health impacts.

self-sufficiency principle The concept that individual nations should take responsibility for waste management in their own country.

sludge A semiliquid mixture of solids, dissolved materials, and water that accumulates in tanks, basins, and sewer pipes from wastewaters or other fluids. Sludge is a component of the waste stream.

solid waste Garbage, refuse, rubbish, sludge, trash, and other discarded material.

source reduction Strategies used to cut down the size, volume, and toxic content of material that is discarded as waste, such as reusing returnable bottles.

source separation Removing recyclable materials from municipal solid waste at the point of origin, considered a front-end process.

thermal treatment The incineration of contaminated soil to destroy organic compounds, one of the most widely used remediation techniques for dealing with hazardous waste.

tipping fee The charges levied to deposit waste at a landfill, waste-to-energy plant, or compost site. Tipping fees are calculated by the ton.

transfer station A central unloading point for municipal solid waste that is partway to its final destination. A transfer station is often the location where local garbage haulers deposit their waste into a large hopper, where it may be compressed to reduce its volume before it is loaded onto a semi-truck trailer, barge, railroad car, or other vehicle to transport it.

vector theory The belief that rats and other vermin were responsible for disease transmission, used as a reason for covering garbage dumps with layers of dirt to keep rodents out.

waste stream The flow of refuse that gets processed through a lengthy and complex system of municipal waste management.

waste-to-energy plant A trash facility that burns waste as fuel through high-temperature combustion, producing electrical energy and steam.

zero waste The elimination of waste before it gets made, at the front end of the production process, rather than treating trash after it already exists, by maximizing recycling; minimizing waste; reducing consumption; and ensuring that products are made to be reused, repaired, or recycled back into nature or the marketplace.

Index

Note: t. indicates table.

About the Author

Jacqueline Vaughn is professor of political science at Northern Arizona University; she received her PhD from the University of California, Berkeley. She is the author of *Environmental Politics: Domestic and Global Dimensions; Green Backlash;* and *George W. Bush's Healthy Forests.*

Printed in the USA
CPSIA information can be obtained
at www.ICGtesting.com
LVHW022104130624
783113LV00004B/162